S0-AAG-762

ECONOMICS AND THE ENVIRONMENT

ECONOMICS AND THE ENVIRONMENT

A Materials Balance Approach

BY

Allen V. Kneese, Robert U. Ayres, Ralph C. d'Arge

RESOURCES FOR THE FUTURE, INC.
1755 Massachusetts Avenue, N.W., Washington, D.C. 20036

Distributed by

The Johns Hopkins Press, Baltimore and London

RESOURCES FOR THE FUTURE, INC.
1755 Massachusetts Avenue, N.W., Washington, D.C. 20036

Resources for the Future is a nonprofit corporation for research and education in the development, conservation, and use of natural resources and the improvement of the quality of the environment. It was established in 1952 with the cooperation of the Ford Foundation. Part of the work of Resources for the Future is carried out by its resident staff; part is supported by grants to universities and other nonprofit organizations. Unless otherwise stated, interpretations and conclusions in RFF publications are those of the authors; the organization takes responsibility for the selection of significant subjects for study, the competence of the researchers, and their freedom of inquiry.

RFF editors: Henry Jarrett, Vera W. Dodds, Nora E. Roots, Tadd Fisher.

This book is one of RFF's studies in the quality of the environment, directed by Allen V. Kneese. Robert U. Ayres is a private consultant in the environmental quality field, and Ralph C. d'Arge is assistant professor of economics at the University of California, Riverside. The manuscript was edited by Pauline Batchelder.

Manufactured in the United States of America

Library of Congress Catalog Card Number 74-134692

ISBN 0-8018-1215-1
Price: $2.50

PREFACE AND ACKNOWLEDGMENTS

This is the first monograph-length report from a new program of research at Resources for the Future, dealing with the management of residuals and of environmental quality. It presents some of the broad concepts on which the program is based and presents some new empirical material as well. It represents an effort to break out of the traditional approach in pollution policy and research, which treats air, water, and solid wastes problems as separate categories. It shows that the external cost phenomena which economists have written much about are not isolated and somewhat freakish aberrations but are inherent in the production and consumption activities of modern economies. A framework is developed for identifying priority research, and some important research areas are described in the last chapter. Several additional articles and monographs are in preparation which will extend the theoretical and empirical work begun here.

We are deeply indebted to many persons for help in the preparation of this monograph. In conversations over the years, Blair Bower has stressed the importance of interdependencies among residuals streams resulting from production and consumption activities and the need for a better framework for analyzing them. He also read the entire manuscript and provided detailed comments. Others currently or recently associated with RFF who read portions of the manuscript or provided other useful inputs are Jerome Delson, Richard Frankel, Mason Gaffney, Robert Haveman, Orris Herfindahl, John Krutilla, and George Löf.

Those outside RFF who have made similar contributions are James Buchanan, Department of Economics, Virginia Polytechnic Institute;

Paul Davidson, Department of Economics, Rutgers—The State University; Robert Dorfman, Department of Economics, Harvard University; Otto Eckstein, Department of Economics, Harvard University; Myrick Freeman, Department of Economics, Bowdoin College; Lester Lave, Graduate School of Industrial Administration, Carnegie-Mellon University; Herbert Mohring, Center for Economic Research, University of Minnesota; Gordon Tullock, Department of Economics, Virginia Polytechnic Institute; and Frank Smith, The Travelers Research Center, Inc.

We are grateful to Erna Belton and Richard McKenna for research assistance.

Finally, we wish to acknowledge that we have made heavy use of papers we have published elsewhere, particularly in the *American Economic Review* and in compendium volumes of the Joint Economic Committee of the United States Congress.

April 1970

CONTENTS

Preface and Acknowledgments v

Chapter I. Perspective

Prelude—Economic Theory and Material Things 1
Externalities and Economic Theory 2
The Flow of Materials 7
Conclusion 13

Chapter II. Material Residuals from Production and Consumption

Introduction 16
Residuals Associated with Energy Conversion 17
Thermal power 19
Transportation 24
Industry and households 28
Residuals from Materials Processing and Industrial Production 30
Potential improvements through treatment and process changes 43
Residuals Associated with Final "Consumption": Households 51
Whole sector—interdependencies 65
Concluding Comment 68
Appendix: Waste Heat and Noise—Energy Residuals 69

CHAPTER III. RESIDUALS, GENERAL EQUILIBRIUM, AND WELFARE ECONOMICS

Introduction 74
Basic Model 76
Inclusion of Externalities 84
Environmental Standards and Decentralized Decision-Making 86
  Some general notes on "second best" problems 86
A Simplified Model of the Economy 89
  Complete or partial immutability and "second best" policy 97
A Recapitulation of Results of Previous Sections 102
Price and Welfare Impacts of Environmental Standards 104

CHAPTER IV. CONCLUSIONS, POLICY, RESEARCH

An Approach to Management of Residuals in a Region 108
  Immediate next steps 111
  Institutional obstacles 111
The Challenge to Research 112
  Introduction 112
  Materials flow 112
  Interindustry models 114
  Residuals handling processes 114
  Projecting residuals 115
  Models of the natural environment 115
  Linkages to affected receptors 116
  Social damages 116
  Institutional arrangements 117
  The challenge to economic theory 117
Postlude—A New Malthusianism? 118

SOURCES FOR CHARTS 120

TABLES

1. Weight of basic materials production in the United States plus net imports, 1963–65 10
2. Fuel consumed in all energy production 18
3. Sources of energy used by various sectors, 1965 18
4. Weight breakdown of various fuels used for energy production 19
5. Thermal power combustion residuals, 1965 21
6. Major gaseous emissions per pound of fuel consumed, 1965 urban driving 25
7. Estimates of residuals from automotive transportation, 1965 26
8. Summary of gaseous residuals from energy conversion, 1965. 30
9. Materials balance for livestock 33
10. Materials balance for processing foods of vegetable origin .. 35
11. Major dissipative uses of industrial organic chemicals, 1963 . 38
12. Major dissipative uses of industrial inorganic chemicals, 1963 38
13. Typical residuals as a per cent of material processed (by weight) 39
14. Estimated reduction of BOD in beet sugar processing, 1949 and 1962 49
15. Selected figures from materials balance for two beet sugar processes 53
16. Gaseous emissions in Sacramento County, 1964 54
17. Gaseous emissions in New York City, 1964 55
18. Average per capita solids and BOD_5 in domestic sewage 56
19. Hypothetical materials balance for humans, 1963 57
20. Total mixed refuse collected in five U.S. cities, 1957–58 60
A–1. Energy costs from new coal-fueled power plants 71
A–2. Differential climatic effects between a city and its environs.. 72

CHARTS

1. Schematic depiction of materials flow 9
2. Residuals from the thermal electric industry 22
3. Production and disposal of products of photosynthesis 32
4. Chemical intermediates in nylon production 40
5. Main processes in a beet sugar plant 46
6. High residual beet sugar production process 47
7. Low residual beet sugar production process 48
8. High residual beet sugar production process—no recirculation 50
9. Low residual beet sugar production process with extensive recycle 52
10. Sewage treatment in the United States (as per cent of population) 59
11. Projection of refuse production trends 61
12. Household residual materials flow (per capita) 66
13. Schematic depiction of a residuals management system 113

ECONOMICS AND THE ENVIRONMENT

Chapter I

PERSPECTIVE

We are now in the middle of a long process of transition in the nature of the image which man has of himself and his environment. Primitive men, and to a large extent also men of the early civilizations, imagined themselves to be living on a virtually illimitable plain. There was almost always, somewhere beyond the known limits of human habitation, and over a very large part of the time that man has been on earth, something like a frontier. That is, there was always some place else to go when things got too difficult, either by reason of the deterioration of the natural environment or a deterioration of the social structure in places where people happened to live. The image of the frontier is probably one of the oldest images of mankind, and it is not surprising that we find it hard to get rid of.[1]

Prelude—Economic Theory and Material Things

The hunter is camped on a great plain with a small fire providing a flickering light and intermittent warmth. Tiny wisps of smoke ascend into a vast, clear night sky. Tomorrow the hunter will move, leaving behind ashes, food scraps, and his own excreta. After ten steps these are lost from sight and smell, probably forever. With them he leaves too his brief speculation about sky and earth, brought on by the loneliness of night, and he peers toward the horizon in search of prey.

The Administrator of the World Environment Control Authority sits at his desk. Along one wall of the huge room are real-time displays,

[1] K. E. Boulding, "The Economics of the Coming Spaceship Earth," in H. Jarrett, ed., *Environmental Quality in a Growing Economy* (Johns Hopkins Press for RFF, 1966).

processed by computer from satellite data, of developing atmospheric and ocean patterns, as well as the flow and quality conditions of the world's great river systems. In an instant, the Administrator can shift from real-time mode to simulation to test the larger effects of changes in emissions of material residuals and heat to water and atmosphere at control points generally corresponding to the locations of the world's great cities and the transport movements among them. In a few seconds the computer displays information in color code for various time periods—hourly, daily, or yearly phases at the Administrator's option. It automatically does this for current steady state and simulated future conditions of emissions, water flow regulation, and atmospheric conditions. Observing a dangerous reddish glow in the eastern Mediterranean, the Administrator dials sub-control station Athens and orders a step-up of removal by the liquid residuals handling plants there. Over northern Europe, the brown smudge of a projected air quality standards violation appears and sub-control point Essen is ordered to take the Ruhr area off sludge incineration for 24 hours but is advised that temporary storage followed by accelerated incineration—but with muffling—after 24 hours will be admissible. The CO_2 simulator now warns the Administrator that another upweller must be brought on line in the Murray Fracture Zone within two years if the internationally agreed balance of CO_2 and oxygen is to be maintained in the atmosphere.

These are extremes of interactions of man with his natural environment. But surely everyone would agree that actuality is much closer to the latter than to the former end of this spectrum. It appears, however, that our concepts of law and economics are somewhere in the middle. These are rooted in the idea that property and private two-party exchange can satisfactorily solve almost all resources allocation problems. Instances of air and water pollution have been regarded as somewhat unusual aberrations which can be satisfactorily treated in an ad hoc and specific way—not as problems of resources allocation on a massive scale.

Externalities and Economic Theory

Economic theory has long recognized in a limited way the existence of "common property" problems and resource misallocations associated with them. It was early appreciated that when property rights to a valuable resource could not be parceled out in such a way that one participant's activities in the use of that resource would leave the others unaffected, except through market exchange, unregulated private exchange would lead to inefficiencies. These inefficiencies were of two types: those associated with "externalities" and those associated with "user costs." The former term refers to certain broader costs (or benefits) of individual action which are not taken into account in deciding to take that action. For instance, the individual crude petroleum producer, pumping from a common pool, has no market incentive to take account of the increased cost imposed on others because of reduced gas pressure resulting from his own pumping. Also, because he cannot be sure that a unit of petroleum he does not exploit now will be available for his later use, acting individually he has no reason to conserve petroleum for later and possibly higher value use. Thus he has no incentive to take account of his user cost. The only limit to his current exploitation is current cost—not the opportunity cost of future returns. Consequently the resource will be exploited at an excessively rapid rate in the absence of some sort of collective action. While these problems were recognized with respect to such resources as petroleum, fisheries, and groundwater, and there was rather sophisticated theorizing with respect to them, still private property and exchange have been regarded as the keystones of an efficient allocation of resources.

To quote the famous welfare economist Pigou:

> When it was urged above, that in certain industries a wrong amount of resources is being invested because the value of the marginal social net product there differs from the value of the marginal private net product, it was tacitly assumed that in the main body of industries these two values are equal.[2]

And Scitovsky, another important student of externalities, after having described his cases two and four, which deal with externalities affecting

[2] A. C. Pigou, *Economics of Welfare* (The Macmillan Co., 1952). Even Baumol who saw externalities as a rather pervasive feature of the economy, tends to discuss external diseconomies like "smoke nuisance" entirely in terms of particular examples. W. J. Baumol, *Welfare Economics and the Theory of the State* (Harvard University Press, 1967). A perspective more like that of the present book is found in K. W. Kapp, *The Social Costs of Private Enterprise* (Harvard University Press, 1950).

consumers and producers respectively, says:

> The second case seems exceptional, because most instances of it can be and usually are eliminated by zoning ordinances and industrial regulations concerned with public health and safety.
>
> . . . .
>
> The fourth case seems unimportant, simply because examples of it seem to be few and exceptional.[3]

It is the main thesis of this book that at least one class of externalities—those associated with the disposal of residuals resulting from modern consumption and production activities—must be viewed quite differently.[4] In reality they are a normal, indeed inevitable, part of these processes. Their economic significance tends to increase as economic development proceeds, and the ability of the natural environment to receive and assimilate them is an important natural resource of

[3] T. Scitovsky, "Two Concepts of External Economies," *The Journal of Political Economy*, Vol. 62 (Apr. 1954), pp. 143–51.

[4] We by no means wish to imply that this is the only important class of externalities associated with production and consumption. Also, we do not wish to imply that there has been a lack of theoretical attention to the externalities problem. In fact, the past few years have seen the publication of several excellent articles which have gone far toward systematizing definitions and illuminating certain policy issues. Of special note are R. H. Coase, "The Problem of Social Cost," *Journal of Law and Economics*, Vol. 3 (Oct. 1960) pp. 1–44; O. A. Davis and A. Whinston, "Externalities, Welfare, and the Theory of Games," *Journal of Political Economy*, Vol. 70 (June 1962), pp. 241–62; J. W. Buchanan and G. Tullock, "Externality," *Economica*, Vol. 29 (Nov. 1962), pp. 371–84; and R. Turvey, "On Divergencies between Social Cost and Private Cost," *Economica*, Vol. 30 (Nov. 1963), pp. 309–13. However, all these contributions deal with externality as a comparatively minor aberration from Pareto optimality in competitive markets and focus upon externalities between two parties. Mishan, after a careful review of the literature, has commented on this as follows: "The form in which external effects have been presented in the literature is that of partial equilibrium analysis; a situation in which a single industry produces an equilibrium output, usually under conditions of perfect competition, some form of intervention being required in order to induce the industry to produce an 'ideal' or 'optimal' output. If the point is not made explicitly, it is tacitly understood that unless the rest of the economy remains organized in conformity with optimum conditions, one runs smack into Second Best problems." E. J. Mishan, "Reflections on Recent Developments in the Concept of External Effects," *The Canadian Journal of Economics and Political Science*, Vol. 31 (Feb. 1965), pp. 1–34.

rapidly increasing value.[5] We suggest below that the common failure to recognize these facts in economic theory may result from viewing the production and consumption processes in a manner which is somewhat at variance with the fundamental physical law of conservation of mass.

Modern welfare economics concludes that if (1) preference orderings of consumers and production functions of producers are independent and their shapes appropriately constrained, (2) consumers maximize utility subject to given income and price parameters, and (3) producers maximize profits subject to these price parameters, then a set of prices exists such that no individual can become better off without making some other individual worse off. For a given distribution of income this is an efficient state. Given certain further assumptions concerning the structure of markets, this "Pareto optimum" can be achieved via a pricing mechanism and voluntary decentralized exchange.

If the capacity of the environment to assimilate residuals is scarce, the decentralized voluntary exchange process cannot be free of uncompensated technological external diseconomies unless (1) all inputs are fully converted into outputs, with no unwanted material and energy residuals along the way,[6] and all final outputs are utterly destroyed in the process of consumption, or (2) property rights are so arranged that all relevant environmental attributes are in private ownership and these rights are exchanged in competitive markets. Neither of these conditions can be expected to hold in an actual economy, and they do not.

Nature does not permit the destruction of matter except by annihilation with antimatter, and the means of disposal of unwanted residuals which maximizes the internal return of decentralized decision units is

[5] That external diseconomies are integrally related to economic development and increasing congestion has been noted in passing in the literature. Mishan has commented: "The attention given to external effects in the recent literature is, I think, fully justified by the unfortunate, albeit inescapable, fact that as societies grow in material wealth, the incidence of these effects grows rapidly. . . ." Mishan, "Reflections." And Buchanan and Tullock have stated that as economic development proceeds, "congestion" tends to replace "cooperation" as the underlying motive force behind collective action, i.e., controlling external diseconomies tends to become more important than cooperation to realize external economies. J. W. Buchanan and G. Tullock, "Public and Private Interaction under Reciprocal Externality," in J. Margolis, ed., *The Public Economy of Urban Communities* (Johns Hopkins Press for RFF, 1965), pp. 52–73.

[6] Or any residuals which occur must be stored on the producer's premises.

by discharge to the environment, principally watercourses and the atmosphere. Water and air are traditionally examples of free goods in economics. But in reality in developed economies they are common property resources of great and increasing value, which present society with important and difficult allocation problems that exchange in private markets cannot solve. These problems loom larger as increased population and industrial production put more pressure on the environment's ability to dilute, chemically degrade, and simply accumulate residuals from production and consumption processes. Only the crudest estimates of present external costs associated with residuals discharge exist, but it would not be surprising if these costs were already in the tens of billions of dollars annually.[7] Moreover, as we shall emphasize again, technological means for processing or purifying one or another type of residuals do not destroy the residuals but only alter their form. Thus, given the level, patterns, and technology of production and consumption, recycle of materials into productive uses or discharge into an alternative medium are the only general operations for protecting a particular environmental medium such as water. Residual problems must be seen in a broad regional or economy-wide context rather than as separate and isolated problems of disposal of gaseous, liquid, solid, and energy waste products.

Frank Knight perhaps provides a key to why these elementary facts have played so small a role in economic theorizing and empirical research.

> The next heading to be mentioned ties up with the question of dimensions from another angle, and relates to the second main error mentioned earlier as connected with taking food and eating as the type of economic

[7] It is interesting to compare this with estimates of the cost of another well-known misallocation of resources that has occupied a central place in economic theory and research. In 1954, Harberger published an estimate of the welfare cost of monopoly which indicated that it amounted to about 0.07 per cent of GNP. A. C. Harberger, "Monopoly and Resources Allocation," *American Economic Review*, Vol. 44 (May 1954), pp. 77–87. In a later study, Schwartzman calculated the allocative cost at only 0.01 per cent of GNP. D. Schwartzman, "The Burden of Monopoly," *Journal of Political Economy*, Vol. 68 (Dec. 1960), pp. 627–30. Leibenstein generalized studies such as these to the statement that "in a great many instances the amount to be gained by increasing allocative efficiency is trivial. . . ." H. Leibenstein, "Allocative Efficiency vs. 'X-Efficiency'," *American Economic Review*, Vol. 56 (June 1966), pp. 392–415. But Leibenstein did not consider the allocative costs associated with environmental pollution.

activity. The basic economic magnitude (value or utility) is service, not good. It is inherently a stream or flow in time. . . .[8]

Standard economic allocation theory is in truth concerned with services. Material objects are merely the vehicles which carry some of these services, and they are exchanged because of consumer preferences for the services associated with their use or because they can help to add value in the manufacturing process. Yet we persist in referring to the "final consumption" of goods as though material objects such as fuels, materials, and finished goods somehow disappear into a void—a practice which was comparatively harmless only so long as air and water were almost literally "free goods."[9] Of course, residuals from both the production and consumption processes remain, and they usually render disservices (like killing fish, increasing the difficulty of water treatment, reducing public health, soiling and deteriorating buildings, etc.) rather than services. These disservices flow to consumers and producers whether they want them or not, and except in unusual cases they cannot control them by engaging in individual exchanges.[10]

The Flow of Materials

To elaborate on these points, we find it useful to view environmental pollution and its control from the perspective of a materials balance problem for the entire economy.[11] The inputs of the system are fuels,

[8] F. H. Knight, *Risk, Uncertainty, and Profit* (New York: A. M. Kelley, 1921). The point was also clearly made by Fisher: "The only true method, in our view, is to regard uniformly as income the *service* of a dwelling to its owner (shelter or money rental), the *service* of a piano (music), and the *service* of food (nourishment) . . ." (emphasis in original). I. Fisher, *Nature of Capital and Income* (New York: A. M. Kelley, 1906).

[9] We are tempted to suggest that the word "consumption" be dropped entirely from the economist's vocabulary as being basically deceptive. It is difficult to think of a suitable substitute, however. At least, the word consumption should not be used in connection with goods, but only with regard to services or flows of "utility."

[10] There is substantial recent literature dealing with the question of under what conditions individual exchanges can optimally control technological external diseconomies. A discussion of this literature, as it relates to waterborne residuals, is found in A. V. Kneese and B. T. Bower, *Managing Water Quality: Economics, Technology, Institutions* (Johns Hopkins Press for RFF, 1968).

[11] As far as we know, the idea of applying materials balance concepts to waste disposal problems was first expressed by Smith. F. A. Smith, "The Economic

foods, and raw materials which are partly converted into final goods and partly become residuals. Except for increases in inventory, final goods also ultimately enter the residuals stream. Thus, goods which are "consumed" really only render certain services. Their material substance remains in existence and must be either reused or discharged to the natural environment.

In an economy which is closed (no imports or exports) and where there is no net accumulation of stocks (plants, equipment, inventories, consumer durables, or buildings), the amount of residuals which is inserted into the natural environment must be approximately equal to the weight of basic fuels, food, and raw materials entering the processing and production system, plus oxygen taken from the atmosphere.[12] This result, while obvious upon reflection, leads to the at first rather surprising corollary that residuals disposal involves a greater tonnage of materials than basic materials processing, although many of the residuals, being gaseous, require comparatively little physical "handling."

Chart 1 shows a materials flow of the type we have in mind and relates it to a broad classification of economic sectors which is convenient for our later discussion and which is generally consistent with the Standard Industrial Classification used by the Census Bureau of the U.S. Department of Commerce. In an open (regional or national) economy, it would be necessary to add flows representing imports and exports. In an economy undergoing stock or capital accumulation, the production of residuals in any given year would be less by that amount than the basic inputs. In the entire U.S. economy, accumulation accounts for about 10–15 per cent of basic annual inputs, mostly in the

---

Theory of Industrial Waste Production and Disposal," draft of a doctoral dissertation, Northwestern University, 1967. We also benefited from an unpublished paper by Joseph Headley in which a pollution "matrix" is suggested. We have also found suggestive references by Boulding to a "spaceship economy" ("The Economics of the Coming Spaceship Earth," pp. 3–14). One of the authors has previously used a similar approach in ecological studies of nutrient interchange among plants and animals; see R. U. Ayres, "Stability of Biosystems in Sea Water," Technical Report No. 142, Hudson Laboratories, Columbia University, 1967. As we note later, residual energy is another major source of external costs and could be analyzed analogously in terms of "energy balance."

[12] To simplify our language, we will not repeat this essential qualification at each opportunity, but assume it applies through the following discussion. In addition, we must include residuals such as NO and NO_2 arising from reactions between components of the air itself but occurring as combustion by-products.

Chart 1. Schematic Depiction of Materials Flow

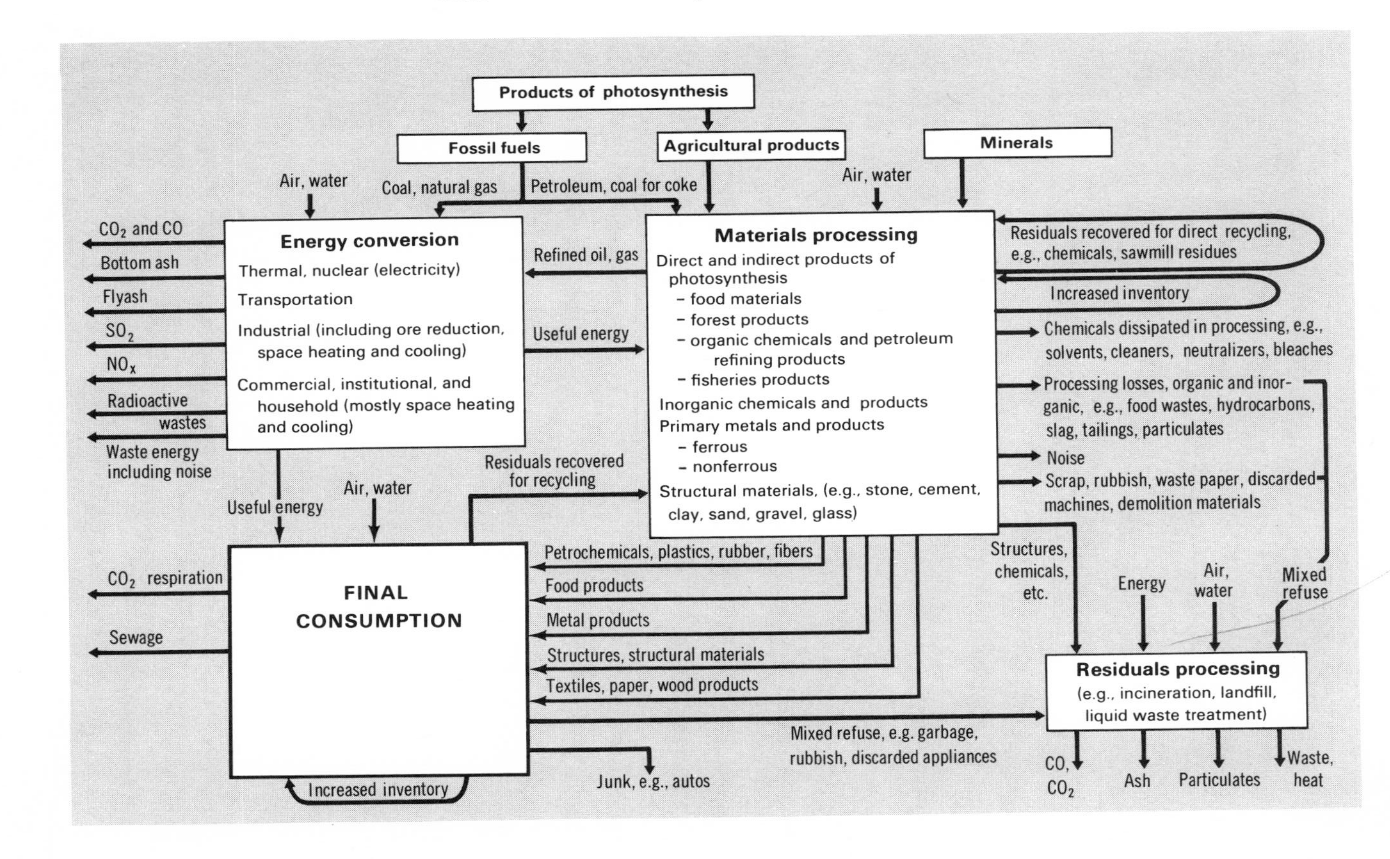

form of construction materials, and there is some net importation of raw and partially processed materials, amounting to 4 or 5 per cent of domestic production. Table 1 shows estimates of the weight of raw materials produced in the United States in several recent years, plus net imports of raw and partially processed materials.

Table 1. Weight of Basic Materials Production in the United States plus Net Imports, 1963–65

(10^6 tons)

Material	1963	1964	1965
Agricultural (incl. fishery and wildlife and forest) products:			
Food and fiber:			
Crops	350	358	364
Livestock and dairy	23	24	23.5
Fishery	2	2	2
Forestry products (85% dry wt. basis):			
Sawlogs	107	116	120
Pulpwood	53	55	56
Other	41	41	42
Total	576	596	607.5
Mineral fuels	1,337	1,399	1,448
Other minerals:			
Iron ore	204	237	245
Other metal ores	161	171	191
Other nonmetals	125	133	149
Total	490	541	585
Grand total[a]	2,261	2,392	2,492

Source: R. U. Ayres and A. V. Kneese, "Environmental Pollution," in *Federal Programs for the Development of Human Resources*, a compendium of papers submitted to the Subcommittee on Economic Progress of the Joint Economic Committee, Congress of the United States, Vol. 2 (U.S. Government Printing Office, 1968).

[a] Excluding construction materials, stone, sand, gravel, and other minerals used for structural purposes, ballast, fillers, insulation, etc. Gangue and mine tailings are also excluded from this total. These materials account for enormous tonnages but undergo essentially no chemical change. Hence, their use is more or less tantamount to physically moving them from one location to another. If this were to be included, there is no logical reason to exclude material shifted in highway cut and fill opera-

Of the "active" inputs,[13] perhaps three-quarters of the overall weight is eventually discharged to the atmosphere as carbon (combined with atmospheric oxygen in the form of CO or CO_2) and hydrogen (combined with atmospheric oxygen as H_2O) under current conditions. This discharge results from combustion of fossil fuels and from animal respiration. Discharge of carbon dioxide can be considered harmless in the short run. There are large "sinks" (in the form of vegetation and large water bodies, mainly the oceans) which reabsorb this gas, although there is some evidence of its net accumulation in the atmosphere. Some experts believe that CO_2 is likely to show a large relative increase—as much as 50 per cent—by the end of the century, possibly giving rise to significant and probably, on balance, adverse weather changes.[14] Thus, continued combustion of fossil fuels at a high rate could produce externalities affecting the entire world. The effects associated with most residuals will normally be more confined, however, usually limited to regional air and water sheds.

The remaining residuals are either gases (like carbon monoxide, nitrogen dioxide, and sulfur dioxide—all potentially harmful even in the short run), dry solids (like rubbish and scrap), or wet solids (like garbage, sewage, and industrial residuals suspended or dissolved in water). In a sense, the dry solids and gases are the irreducible, limiting forms of residuals. By the application of appropriate equipment and energy, most undesirable substances can, in principle, be removed from water and air streams,[15] but what is left must be disposed of in solid

tions, harbor dredging, land-fill, plowing, and even silt moved by rivers. Since a line must be drawn somewhere, we chose to draw it as indicated above.

[13] See footnote to Table 1.

[14] *Implications of Rising Carbon Dioxide Content of the Atmosphere*, Conservation Foundation, New York, 1963. There is strong evidence that discharge of residuals has already affected the climate of individual cities. W. P. Lowry, "The Climate of Cities," *Scientific American*, Vol. 217 (Aug. 1967), pp. 15–23. The other side of the coin of CO_2 production is oxygen consumption, which some experts also regard as a potential problem. In the United States in 1965 about 1.3 billion tons of fossil fuels were consumed for all purposes along with 2.74 billion tons of atmospheric oxygen, to yield 3.77 billion tons of CO_2 plus immense tonnages of assorted by-products. In comparison, human respiration requires about 60 million tons of atmospheric oxygen (for the population of the United States) and produces 98 million tons of CO_2. Based on biomass, the sum total of all animals—mainly cattle—would require less than five times as much oxygen as the human population alone, or somewhere in the neighborhood of 0.3 billion tons, and produces 0.5 billion tons of CO_2 at most.

[15] Except CO_2, which may be harmful in the long run, as noted.

form, transformed, or reused. Looking at the matter in this way clearly reveals a primary interdependence between the various residuals streams which casts into doubt the traditional classification of air, water, and land pollution as individual categories for purposes of planning and control policy.[16]

Material residuals do not necessarily have to be discharged to the environment. In many instances it is possible to recycle them back into the productive system. The materials balance view underlines the fact that the throughput of new materials necessary to maintain a given level of production and consumption decreases as the technical efficiency of energy conversion and materials utilization increases. Similarly, other things being equal, the longer cars, buildings, machinery, and other durables remain in service, the fewer new materials are required to compensate for loss, wear, and obsolescence—although the use of old or worn machinery (e.g., automobiles) tends to increase other residuals problems. Technically efficient combustion of (desulfurized) fossil fuels would leave only water, ash, and carbon dioxide as residuals, while nuclear energy conversion need leave only negligible quantities of material residuals, although pollution from discharge of heat—an energy residual—and of radiative materials cannot be dismissed by any means.

Given the population, industrial production, and transport services in an economy (a regional rather than a national economy would normally be the relevant unit), it is possible to visualize combinations of social policy which could lead to quite different relative burdens placed on the various residuals-receiving environmental media. And, given the possibilities for recycle and less residual-generating production processes, a lighter overall burden can be placed upon the environment as a whole. To take one extreme, a region which went in heavily for electric space heating and wet scrubbing of stack gases (from steam plants and industries), which ground up its garbage and delivered it to the sewers and then discharged the raw sewage to watercourses, would pro-

[16] Water pollution on the federal level is presently the province of the Federal Water Pollution Control Administration (Department of the Interior); air pollution, of the National Air Pollution Control Administration (Department of Health, Education, and Welfare); and solids, of the Bureau of Solid Wastes Management (also DHEW). Similar divisions are usually present at the state and local levels.

tect its air resources to an exceptional degree. But this would come at the sacrifice of placing a heavy residuals load upon water resources. On the other hand, a region which treated municipal and industrial liquid residuals streams to a high level but relied heavily on the incineration of sludges and solid residuals would protect its water and land resources but at the expense of discharging residuals predominantly to the air. Finally, a region which practiced high-level recovery and recycle of residuals and fostered low residuals production processes to a far-reaching extent in each of the economic sectors might discharge very few residuals to any of the environmental media.

Further complexities are added by the fact that sometimes it is possible to modify an environmental medium through investment in control facilities so as to improve its assimilative capacity. The clearest, but far from only, example is with respect to watercourses where reservoir or groundwater storage can be used to augment low river flows which ordinarily are associated with critical water quality levels (high external cost situations).[17] Thus, internalization of external costs associated with particular discharges by means of other restrictions, even if done perfectly, cannot guarantee Pareto optimality. Investments involving public good aspects must enter into an optimal solution.[18]

Conclusion

Air and water used to be the economist's favorite examples of "free goods" (goods so abundant that their marginal value to any user, or potential user, is zero). It was historically fortunate for conventional economic theorizing about the efficiency of market processes that this was approximately true in fact. These media served the function of "infinite sinks" for most of the residuals which are, as the materials balance view so clearly shows, an inevitable accompaniment of production and consumption activities.

What is appearing now, however, is a vast asymmetry in the ade-

[17] Careful empirical work has shown that this technique can fit efficiently into water quality management systems. See R. K. Davis, *The Range of Choice in Water Management* (Johns Hopkins Press for RFF, 1968).

[18] A discussion of the theory of such public investments with respect to water quality management is found in J. H. Boyd, "Collective Facilities in Water Quality Management," appendix to Kneese and Bower, *Managing Water Quality*.

quacy of our property institutions (which, of course, underlie all private exchange) to handle resources allocation problems. On the one hand, in the production of basic natural resources commodities, property institutions with some controls and adjustments, in general, serve quite well to lead production into highest productivity channels now and in the future. On the other hand, the flow of residuals back to the environment is heavily weighted to media where private property institutions can function imperfectly, if at all. Once these media become overloaded on a significant scale, they are free goods no more but, rather, *natural resources* of ever increasing value as economic development proceeds.

To recapitulate briefly our main points so far:

(1) Technological external diseconomies are not freakish anomalies in the processes of production and consumption but inherent and normal parts of them.

(2) These external diseconomies are quantitatively negligible in a low-population economically undeveloped setting, but they become progressively (nonlinearly) more important as the population rises and the level of output increases (i.e., as the natural reservoirs of dilution and assimilative capacity become exhausted).[19]

(3) They cannot be properly dealt with by considering environmental media such as air and water in isolation.

(4) Isolated and ad hoc taxes and other restrictions are not sufficient for their optimum control, although such policy instruments are essential elements in a more systematic and coherent program of environmental quality management.

(5) Public investment programs, particularly including transportation systems, sewage disposal, and river flow regulation, are intimately related to the amounts and effects of residuals and must be planned in light of them.

In view of this it is important to develop not only improved measures of the external costs resulting from differing concentrations and durations of residuals in the environment, but more systematic methods

[19] Externalities associated with residuals discharge may appear at certain threshold values which are relevant only at some stage of economic development and industrial and population concentrations. This may account for their general treatment as "exceptional" cases in the economics literature. These threshold values truly would be exceptional cases for less developed agrarian economies.

for projecting emissions of external cost producing residuals, technical and economic tradeoffs among them, and the effects of recycle on environmental quality.

In the following chapters we hope to make a modest beginning in this direction. In the chapter immediately following, we elaborate on the materials balance concept and present numerous suggestive empirical results.

In the third chapter we display a formal mathematical model which weds the materials balance concept to the concept of a general equilibrium in exchange. This model is of theoretical interest and also suggests how the more simplified versions of economic general interdependency models might be used to provide more coherent and accurate projections of residuals discharges as a function of the final products of an economy. We also undertake an examination of whether it is possible to conceive of coherent and consistent environmental standards in the presence of pervasive and interdependent externalities resulting from residuals discharge. The welfare implications of such standards are investigated and some comments made about the deficiencies of presently available economic models in connection with such questions.

In the final chapter we undertake to draw some practical conclusions from our analysis and suggest some important directions for future research.

Chapter II

MATERIAL RESIDUALS FROM PRODUCTION AND CONSUMPTION

Introduction

We have introduced the concept of material balance for the economy as a whole and pointed to a few of its implications for economics and public policy. Now we apply this concept to an analysis of material residuals from major sectors of the national economy. While numerous empirical results are presented, these must be understood to have primarily an illustrative and propaedeutic value. This is because almost all of them relate to national aggregates rather than to the regions in which actual planning, design, and operation of management systems would have to be concentrated. Even so, these results do yield considerable insight into how gaseous, liquid, and solid residuals are generated by production and consumption activities in the economy and why it is important to understand their interrelationships. Also, this chapter provides us with an opportunity to introduce some of the major residuals control technologies. The large economic sectors we analyze are energy conversion, processing, and consumption. Materials inputs and outputs of these sectors were depicted schematically in Chart 1 of the previous chapter, which can be regarded as providing an outline for the present discussion.

A parallel analysis of energy balances and energy residuals could be developed. We have not done this—except to a very limited extent in the appendix to this chapter. In concept, extension of the analysis to energy balances and energy residuals is straightforward, but it might well involve substantial problems in application.

In reading this chapter, it would be well to bear in mind that the technology available for handling and recycling residuals is primitive in relation to that for extraction, production, and distribution activities (i.e., the residuals generating activities). A main reason can be inferred from the common property concept introduced in the preceding chapter. A static allocative effect of the discharge of residuals to unpriced common property environments is that presently available technology is not used to an optimal extent in their management and control. A dynamic version of this is that the market economy provides no automatic incentives for the development of new technology—as it does for extraction, processing, and distribution activities.[1]

Residuals Associated with Energy Conversion

The energy conversion sector, as a whole, obtained 4 per cent output from hydroelectric generators and 0.1 per cent from nuclear fuel in 1965. The remainder was derived from fossil fuels as shown in Table 2.

In 1965, electric utilities consumed 20 per cent of all primary energy, 24 per cent went to transportation (mainly as gasoline), 32 per cent was used in industry, and 21 per cent was used in households and commercial establishments. The overall breakdown of energy consumption, by source, of the various sectors is shown in Table 3. Coal dominates the electric power generating field and plays an important role in industry, especially in smelting of ferrous metals; petroleum even more

[1] After the present chapter was written, two rather broad studies of waste disposal problems were published. The interested reader is referred to them for good discussions and, in some instances, slightly more recent or complete data on some of the residuals we consider in this chapter. They are: *Cleaning Our Environment: The Chemical Basis for Action* (Washington, D.C.: American Chemical Society, 1969), and *Solid Waste Management: A Comprehensive Assessment of Solid Waste Problems, Practices, and Needs,* prepared by the Office of Science and Technology, Executive Office of the President (Washington, D.C., May 1969).

In general, we have concerned ourselves with basic mining and agricultural products after they have been extracted from the land and entered the processing sector. Doing this is more practical from a data standpoint than analyzing the complete process including extraction, and it does focus on the materials flow and residuals generation which most directly affect concentrations of people. However, it does leave aside important residuals which often have external effects, like mine tailings, acid mine drainage, and burning of field wastes, for example.

Table 2. Fuel Consumed in All Energy Production

Fuel source	Contribution to total primary energy	1965 consumption as fuel
	(Per cent)	(Million tons)
Coal	23	465
Petroleum and natural gas liquids	43	503
Natural gas, dry[a]	30	337

SOURCE: Figures from W. A. Vogely and W. E. Morrison, "Patterns of Energy Consumption in the United States, 1947–65 and 1980 Projected," *Transactions* of the 15th Sectional Meeting of the World Power Conference, Oct. 1966 (Tokyo, Japan: Japanese National Committee of the World Power Conference, 1966).

[a] Assuming an average molecular weight of 16 (methane, CH_4), whence 16 grams (1 mole) occupies 22.4 liters; thus, 337 million tons of natural gas is equivalent to 15.2 trillion cubic feet.

Table 3. Sources of Energy Used by Various Sectors, 1965

(Per cent)

Energy source	Utility	Transportation	Industry	Households and commercial
Coal	55	neg.	29.4	0.4
Petroleum	6	99.85	22.8	38.8
Natural gas	22	neg.	39.3	42.8
Utility electricity	([a])	0.15	8.5	14.4
Other	17	neg.	neg.	3.6
Total	100	100.0	100.0	100.0

SOURCE: Vogely and Morrison, "Patterns of Energy Consumption."
neg. Negligible.
[a] Not applicable.

heavily dominates the transportation area. These two fuels also cause the most serious residuals problems, as will be seen later.

To discuss the pattern of utilization of mineral fuels in the U.S. economy, it is necessary to analyze them in terms of their primary constituents—carbon, hydrogen, and sulfur—as indicated in Table 4. We can then follow these elements through a sequence of chemical transmutations without being unduly concerned with the exact physical form or chemical combination at each stage.

Table 4. Weight Breakdown of Various Fuels Used for Energy Production

Fuels used	Per cent carbon[a] (ex. CO_2)	Per cent hydrogen[a]	Per cent sulfur[b]	Per cent other (ash, H_2O, CO_2, etc.)
Anthracite coal, 16.7 per cent ash	76.0	2.6	0.6–0.8	~20.0
Bituminous coal, 7.5 per cent ash (after washing at mine)	75.0	5.0	2.0	18.0
Electric utilities coal, 10 per cent ash (after washing at mine)	75.0	4.5	2.5	18.0
Natural gas, unprocessed, 93 per cent CH_4 + 4 per cent C_2H_6	75.0	24.0	0	1.0
Gasoline (motor fuel), assumed C_8H_{18}	84.0	16.0	0	<0.5
Distillate fuel oil (grades 1 to 4)	85.0[c]	14.5[c]	neg.	0.5
Residual fuel oil (grades 5 to 6)	85.7	11.5	2.0	0.8
Residual for utilities	85.4	11.4	2.5	0.7

neg. Negligible.

[a] Figures on anthracite and bituminous coal from *Combustion Engineering*, 1966 ed., Table 13–2. Figures on utility coal, residual oil, and natural gas from *Costs of Large Fossil Fuel Fired Power Plants*, Jackson & Moreland, Inc., Boston, Mass., Apr. 1966.

[b] These figures from "Sulfur in U.S. Coals," *Coal Age*, Dec. 1955, pp. 78–79, and from E. A. Fohrman, J. H. Ludwig, and B. J. Steigerwald, "Coal Utilization and Atmospheric Pollution," *Coal*, Apr. 1965, pp. 5–6.

[c] Estimated.

Thermal Power[2]

As noted above, over half the utility electric power produced in 1965 used coal as the primary energy source. In 1965 about 250×10^6 tons out of a total U.S. domestic supply of 465×10^6 tons were used for this purpose.[3] Most of the remaining electric energy was produced with hydropower and natural gas. About half the natural gas used by electrical utilities is used in the South Central region; the rest is spread out. Combustion of natural gas produces comparatively small amounts of potentially harmful residuals.[4] Some 16.5×10^6 tons of (high sulfur)

[2] Much of the information for this section was obtained from an unpublished research report by Richard J. Frankel, formerly of Resources for the Future, Inc., August, 1967.

[3] About 100×10^6 tons each were used by the steel industry and other industries.

[4] Use of natural gas does cause some oxides of nitrogen to be produced as secondary products. During combustion, oxygen and nitrogen gas combine to form

residual oil were also burned by electric powerplants, mostly on the East Coast.

A variety of residuals which can result in external costs are associated with the use of coal. Among them are turbid waste waters from coal preparation at the mine, losses as soot or dust during transport, flyash (fine inorganic particulates in flue gases), bottom ash, and gaseous stack emissions (primarily CO_2, sulfur oxides, and oxides of nitrogen or NO_X)—and, if flue gases are scrubbed, waterborne residuals result from this process. Indeed, coal and residual oil burning thermal plants contribute major fractions of the total amounts of sulfur dioxide (~50 per cent), NO_X (~50 per cent), and particulates (~25 per cent) emitted to the atmosphere in the United States. In the case of particulates, this remains true despite the fact that major control efforts have already been introduced. A summary of the residuals from nonnuclear thermal power production is shown in Table 5.

Chart 2 shows several alternative process sequences giving rise to varying amounts of residuals. A high sulfur coal burning operation with little or no control or recycling of residuals back into the productive system can give rise to high levels of residuals generation at various stages and potentially large external costs. On the other hand, we also give a schematic representation of a sequence which uses high sulfur coal but produces few residuals of a directly harmful kind. Other process sequences might be sketched, but the one shown seems a reasonable one for achieving low residuals, based on presently available or reasonably foreseeable technology. No control and reuse system as complete as this, however, is presently employed at any U.S. thermal power plant.

In this process sequence water used for coal preparation at the mine is ponded and the resulting sludge is used for landfill. Next, ash and pyrites (metallic sulfide crystals) are removed from the coal by mechanical means at the mine. (The pyrite can be used to produce commercial sulfuric acid.) The coal is then transported by unit train or slurry pipeline, to reduce losses in transportation. After combustion,

---

nitric oxide (NO), which will react with more oxygen to form nitrogen dioxide (NO_2) and other nitrogen oxides (generally written NO_X) which contributes to Los Angeles-type smog. It has recently been discovered, apparently by accident, that a two stage combustion process can greatly reduce NO_X from natural gas and oil fired plants, but this technology has not been developed for coal fired plants.

Table 5. Thermal Power Combustion Residuals, 1965

	Weight breakdown (see Table 3)				
Fuels used	Total wt. x 10^6 tons	Weight carbon x 10^6 tons	Weight hydrogen x 10^6 tons	Weight sulfur x 10^6 tons	Weight ash x 10^6 tons
Total coal:	251	168	11.3	6.4	25
Anthracite coal	2				
Bituminous coal	249				
Natural gas	51	38	12.5	0.0	0.0
Residual oil[a]	16.5	14.2	1.9	0.4	0.0
Total	318.5	220.2	25.7	6.8	25

Residuals (x 10^6 tons): CO_2 807
H_2O 231
SO_2 . . . . 13.6[b]
Ash (total) 25
Flyash not collected . . . . 2.4[c]
NO_X . . . . 3.7[d]

[a] 16.5 x 10^6 tons = 101 x 10^6 bbls.

[b] The Public Health Service estimates 12 x 10^6 tons of SO_2 emitted into the atmosphere by electric utilities, which might suggest that 1.7 x 10^6 tons is being removed from stack gases at present; however, it appears that SO_2 removal at present is actually much less than this.

[c] Data from Frankel, unpublished research report. Based on an industry-wide collection efficiency of 86.5% for flyash in the stack, and allowing about 20% for bottom ash.

[d] Assuming 28.4 lbs. NO_X produced per ton of oil burned (Los Angeles). If carbon is the controlling factor (as seems to be the case), then we would expect 33.3 lbs. NO_X per ton of contained carbon or 3.7 x 10^6 tons NO_X on a nationwide basis.

the CO_2 is discharged harmlessly (at least in the short run) into the ambient air. Up to 90 per cent of the sulfur dioxide is removed from the stack gases by a dry removal process, and the recovered sulfur also is saleable.[5] Flyash is removed up to 99 per cent by electrostatic precipitators and used for landfill or construction materials (light aggregate). The overall amount of nonrecycled waste (excluding CO_2) from such a high level residuals control process (even giving special weight to the potentially most harmful residuals) is perhaps less than 10 per

[5] Dry removal processes need some further development before they can be economically substituted for wet scrubbing, which is effective in removing SO_2 and flyash but which leaves a large waterborne residual.

Chart 2. Residuals from the Thermal Electric Industry

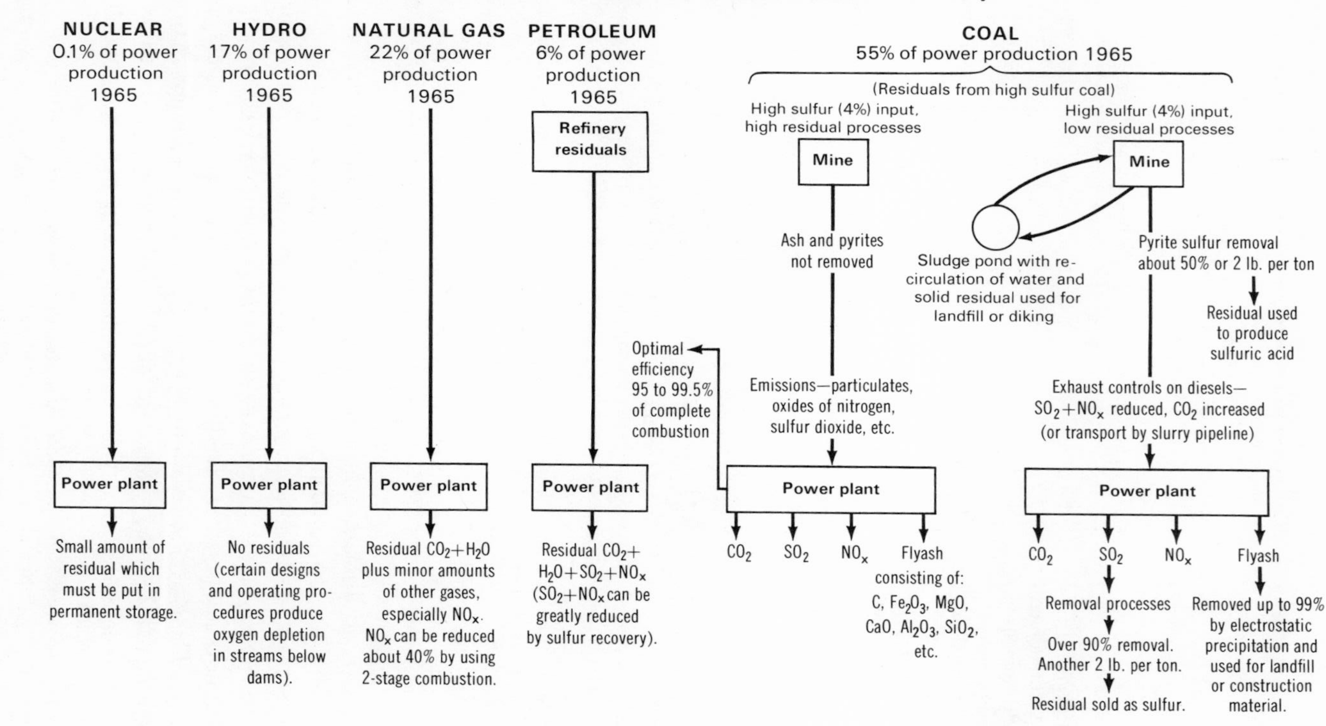

cent of that resulting from the less strictly controlled process. Moreover, since virtually all flyash and sulfur are productively reused, potential losses in other production processes are short-circuited (for example, less elementary sulfur and pumice need be mined).

However, the dry SO_2 removal process does not remove NO_X from the gas stream nor (as was mentioned in footnote 4) is two stage combustion presently applicable to conventional coal fired furnaces. Wet scrubbing will remove some NO_X as well as SO_2, but again the result is a waterborne waste stream. The interdependencies and tradeoffs between various waste streams which must be considered thus are clearly illustrated.

An obvious substitute for emissions control for "dirty" fuels is to use a "clean" fuel like natural gas or to convert to atomic energy, and a trend toward using atomic energy is now clearly evident. The amount of residuals produced by nuclear fuels is quantitatively very small compared with that of fossil fuels, since energy released per unit of material throughput is vastly larger. There are no significant discharges of residuals to the atmosphere from nuclear plants, but some of the liquid or solid residuals (small amounts) are so radioactive that they must be put in permanent storage. The full ramifications of this apparently have not been fully explored. Additionally, present generation of nuclear power plants results in 40–50 per cent more waste heat per kwh, and substantial amounts of low level radiation are not containable. Controversy over allowable amounts is now rife.

Despite the growth of atomic energy, the use of fossil fuels for power generation will continue to increase until at least 1975 or 1980. During this period the use of natural gas also is likely to increase. Pipeline capacity could be augmented with a view to shifting gas gradually to household and commercial users as those demands expand and electric power generation shifts to atomic energy.[6]

So far we have focused on overall emissions reduction, but it may also be possible to use systems for making better use of natural assimilative capacity. For example, high stacks (very cheap compared to the cost of reducing SO_2 emissions) help disperse gases and could in large systems perhaps be combined with systematic load shifting from plant

[6] It is also possible that shifts could be made to low sulfur coal, but because of its limited supply and high value to the steel industry it is doubtful that much of this will be available to the eastern utilities which now use high sulfur coal.

to plant to take account of atmospheric conditions.[7] These kinds of alternatives are discussed further in the final chapter, where we address the matter of regional environmental management systems.

Transportation

In 1965 the transportation sector accounted for 24 per cent of all primary energy produced in the United States, and all but an insignificant percentage was accounted for by petroleum products; in 1965 the sector consumed 308×10^6 tons as fuel,[8] plus substantial quantities of additives, notably 2×10^6 tons of lead.

In the United States, transportation is overwhelmingly automotive. In 1965, 82 per cent of all workers used private automobiles en route to and from work and a majority of the rest traveled on buses. Intercity travel is even more heavily dominated by motor vehicles: about 90 per cent of intercity passenger miles were accounted for by private cars and a further 2.5 per cent by motor coaches. The movement of freight is still largely by railroads, but trucks were the second most important mode, with about 0.25 per cent of all freight ton-miles in 1965—and a virtual monopoly of local distribution of goods. Domestic airlines accounted for just under 6 per cent of passenger miles and 0.1 per cent of freight ton-miles.[9]

In the mid-sixties, gasoline motor fuel accounted for 200 million tons out of a total of 300-plus million tons consumed by the transportation sector as a whole. Diesel trucks, buses, and off-highway equipment used 24 million tons; aviation—a rapidly increasing sector[10]—required 39 million tons; and marine and railroad uses accounted for 28 million tons. Some fuel attributed to the transportation sector was used for nontransportation purposes such as space heating of stations and ter-

[7] The possibility of load dispatching to take account of differential atmospheric conditions is discussed in J. K. Delson, "Choices for Electrical Utilities in the Control of Air Pollution," mimeo., Resources for the Future, July 18, 1967.

[8] W. A. Vogely and W. E. Morrison, "Patterns of Energy Consumption in the United States, 1947–65 and 1980 Projected," *Transactions* of the 15th Sectional Meeting of the World Power Conference, Oct. 1966 (Tokyo, Japan: Japanese National Committee of the World Power Conference, 1966).

[9] From "Automobile Facts and Figures," 1966 ed., published by the Automobile Manufacturers Association.

[10] In 1955, airlines accounted for 3.42 per cent of domestic intercity passenger-miles. By 1965 the figure had risen to 5.96 per cent.

minals. Fuel other than gasoline used for *land* transportation, which contributes the bulk of air pollution, amounted to only 38 million tons, mainly diesel. Hence it seems reasonable to focus on the emissions from internal combustion engines. A summary of emissions per unit of fuel consumed is shown in Table 6.

Using the calculated coefficients in Table 6 and the total gasoline consumption in the transportation sector, one obtains the results shown in Table 7. (No allowance is made for aircraft or diesel engines.)

Potential control techniques have been discussed at great length in recent years. Basically these techniques consist of:

(1) Recycling the blow-by gases which formerly escaped from the cylinders past the rings and were formerly released from a vent under the engine (this change has been made since 1963 on all U.S. cars; it eliminates about 20 per cent of the unburned hydrocarbons and carbon monoxide).

(2) Improving the efficiency of the combustion process by improved

Table 6. Major Gaseous Emissions Per Pound of Fuel Consumed, 1965 Urban Driving

(Pounds)

Emission	CO	HC	NO_X	Particulates	SO_2 H_2S
Spark ignition i.c.e.					
(Otto cycle)[a,b]	0.377	0.082	0.0185	0.002	0.0015
Diesel engine[b]	0.0085	0.019	0.032	0.016	0.006
Gas turbine					
(Rover 2S–140)[c]	0.02	neg.	0.0003	neg.	0.002
External combustion					
steam (Williams)[d]	0.0075	0.0016	0.001	neg.	0.006
	(500)	(20)	(70)		

[a] Assuming an average route speed of 24 mph, 40% of HC emissions accounted for by exhaust, the remainder being due to evaporative losses from carburetor and gas tank.

[b] Data from "Compilation of Air Pollutant Emission Factors." U.S. Public Health Service Publication No. 999-AP-42 (1968).

[c] Data calculated from Noel Penny's "Rover Case History of Small Gas Turbines." Paper No. 634A.

[d] Calculated by assuming volumetric emissions shown in parentheses (in ppm) 15 lbs. of air/lb. of fuel, and typical molecular weights. Sulfur emissions assumed to be the same as diesel, with similar fuel. If external combustion engines use more than 15 lbs. of air/lb. of fuel, coefficients should be increased by the corresponding ratio.

Table 7. Estimates of Residuals from Automotive Transportation, 1965

Gaseous residuals	Million tons
Carbon monoxide (CO)	66
Unburned hydrocarbons (HC)	16
Oxides of nitrogen (NO_X)	6
Lead compounds	0.2
Oxides of sulfur (SO_X)	0.2

SOURCE: "The Sources of Air Pollution and Their Control," Public Health Service Publication No. 1548 (U.S. Government Printing Office, 1966).

carburetion (or fuel injection). This is a matter of more accurately matching the fuel and air intake to the instantaneous demands of the engine, and particularly preventing excessively rich mixtures. Evaporation losses from the carburetor can also be considerably reduced by simple redesign.[11]

(3) Providing for more complete consumption of unburned components in exhaust gases, either in the manifold or in the tailpipe, by some type of afterburner. The simplest method is to introduce excess air into the hot manifold, where unburned gases will have an additional opportunity to burn completely. Some version of this method is now in use on all U.S. cars manufactured since 1968. Catalytic mufflers and other add-on devices can also reduce exhaust emissions, although catalysts tend to be poisoned by the lead in the combustion products. Hence the latter approach is not much in favor at present. However, the former method has the unfortunate by-product of increasing NO_X emissions. As yet it is unclear how NO_X standards will be met when they go into effect in California in the early seventies.

(4) Reducing evaporation losses from the gas tank by various means, such as an absorbent charcoal buffer. It is theoretically possible that by a combination of these means about 90 per cent of the emissions from a typical gasoline internal combustion engine may ultimately be eliminated at an acceptable cost (less than $100 or so).

[11] Conversion from gasoline to natural gas as fuel when vehicles are operated in densely populated areas has recently been proposed. Technical feasibility of such conversions has been demonstrated by Pacific Gas Co. in Los Angeles. A loss of power occurs but operating costs are reduced and smog-contributing emissions substantially reduced.

Engines may even achieve somewhat better results than this, while they are comparatively new, but the level of control tends to be degraded with age and wear. The problem of keeping emissions low throughout the life of the car, under conditions of little or poor maintenance, remains a very difficult one. Meanwhile, if emissions control is effective primarily during the first half of a vehicle's life, the net gain by 1980 would ultimately be something like 50 per cent as compared with today's emission levels. However, an expected rapid increase in the number of cars on the roads would nullify most of this gain in the years after that.

To "roll back the clock" more than a few years, more radical approaches would seem to be required, such as:

(1) Greater emphasis on mass transportation and rail—rather than highway—transportation of goods. As an example of what can be achieved by switching automotive traffic to rails, each 100 million passenger miles on electrified mass transit results in saving 16,000 tons of gasoline, which under present conditions means a net decrease of 6,000 tons of carbon monoxide, 1,300 tons of vaporized hydrocarbons, and 200 tons of oxides of nitrogen. Assuming the power is produced in thermal plants, there will be compensating increase in production of oxides of sulfur (assuming fuel with a 2.5 per cent sulfur content). This shift might be accomplished in part by public subsidy of conventional or unconventional mass transit systems, tax or other disincentives to drive or park private cars in areas where alternatives exist, and construction of convenient (automated?) parking facilities near transit termini. Automated comprehensive underground goods distribution systems utilizing electric minirails, pneumatic tubes, and/or moving belts may find a place in densely populated areas.

(2) A (partial?) switch to alternative power sources for vehicles, including gas turbines or turboelectric hybrids for large buses and articulated trucks, and external combustion (e.g., steam) engines or battery fuel cell electric propulsion for automobiles and taxis. Remote (i.e., wired or radiated) external power sources in a dual mode (e.g., road-rail) configuration are also conceivable possibilities for private vehicles in the longer time frame.[12]

In general, these latter technological alternatives can provide trans-

[12] See a forthcoming RFF sponsored study by Robert U. Ayres on technological alternatives to the internal combustion engine; also, U.S. Department of Commerce, *The Automobile and Air Pollution: A Program for Progress*, a compilation

portation services with less throughput of materials and fuel than conventional motor vehicles and/or involve combustion of fuels in central power plants where it is more efficient and residuals can be more efficiently handled. They can be encouraged by public sponsorship of research, public investment, tax incentives such as faster writeoffs, effluent taxes on emissions, enforcement of air quality standards, and directed purchase of appropriate vehicles for the use of government agencies. Thus the Post Office alone would provide a sizable market for an electric truck, and city, state, and federal agencies could provide an initial market for a steam engine or battery fuel cell (electric) vehicle. Once the economies offered by such vehicles are proven, fleet owners such as rent-a-car concerns and taxi companies, and finally private individuals, might be induced to switch.

The externalities caused by electric power production and by automotive transportation make an interesting contrast. Electric power producers are acknowledged to be *utilities* subject to control and supervision by public agencies; hence the regulation route mentioned earlier is quite likely to be successful in the long run as a means of enforcing due consideration of the external costs of operation.

On the other hand, although railroads, airlines, and interstate trucking companies are regulated as utilities, the private automobile, which accounts for the great bulk of all transportation, is treated not as a utility but as a convenience. Neither the manufacturers nor the owners have, until very recently, been subject to any systematic regulation representing the public interest. Scattered state and federal legislation now exists covering such topics as auto safety, compulsory insurance, and gaseous emissions; but these functions are not unified under a single agency, nor are they likely to be in the immediate future. This fragmentation of governmental authority, coupled with the enormous economic power of the automotive industry, makes a successful attack on the problem of automobile-caused deterioration in environmental quality difficult.

Industry and Households

As noted previously, coal is an important source of energy in industry, particularly in the metallurgical field and in the manufacture of

---

of sub-panel reports to the Advisory Panel on Electrically Powered Vehicles (Dec. 1967), Vols. I and II.

lime and portland cement. Thus in 1965 the industrial sector consumed 188×10^6 tons of coal. Of this quantity, 96×10^6 tons were first carbonized to yield 77×10^6 tons of coke plus coal-gas and about 4×10^6 tons of coal tar derivatives such as benzene, toluene, xyline, naphthalene, and creosote, which by-products were mostly used as raw materials for the manufacture of organic chemicals. The best quality low sulfur (~1 per cent) bituminous coals normally are used for coking.[13] Thus the average sulfur content of all industrial coal (2 per cent) is less than the average for utility coal (2.5 per cent), as shown in Table 4. This implies a total SO_2 emission from coal burned in industry of 7.4×10^6 tons in 1965.[14] Total particulates (flyash) produced by industrial and coking coal are 14×10^6 tons, based on the average 7.5 per cent ash content shown in Table 4. Assuming an average collection efficiency of 62 per cent implies residual particulate emissions on the order of 5.3×10^6 tons.[15] The Public Health Service estimates a total of 6×10^6 tons of particulate matter from all industrial operations.[16] Although we have no quantitative basis of verification, the same source also estimates that 2×10^6 tons of NO_X and 2×10^6 tons of carbon monoxide are produced annually by industry in the United States. Further discussion of energy conversion residuals production and controls in materials processing and manufacturing operations is best reserved for the next section, where the industrial sector is analyzed explicitly in more detail.

Space heating in industry and households is the other major source of demand for energy. The fuels used in this application are mainly distillate oils (~177×10^6 tons) and natural gas (286×10^6 tons), both of which are comparatively clean. About 26×10^6 tons of coal are still used in the household sector, although this use of coal is decreasing rapidly. The latter contributes on the order of 1×10^6 tons of SO_2 and —because of the virtually complete lack of flyash collection in small heating plants—probably 1.7×10^6 tons of particulates. The ash content (0.5 per cent) of fuel oil probably contributes a further 1×10^6

[13] These low sulfur coals, mainly from western Pennsylvania and West Virginia, sell for much higher prices than utility coal, and supplies are largely preempted by long term contracts (or are in "captive" mines).

[14] The Public Health Service estimated 9×10^6 tons of SO_2, which presumably includes SO_2 from other sources such a sulfide ore treatment.

[15] Frankel, unpublished research report.

[16] This is certainly too low if additional contributions from mineral ore beneficiation (particularly copper) and phosphatic fertilizer plants are considered.

Table 8. Summary of Gaseous Residuals from Energy Conversion, 1965
(Million tons)

Energy user	Carbon monoxide (CO)	Hydro-carbons (HC)	Sulfur dioxide (SO_2)	Oxides of nitrogen (NO_X)	Particu-lates
Utility power	1	neg.	13.6	3.7	2.4
Industry and households	5	neg.	8.4	7.0	7.0
Transportation	66	12	0.4	6.0	0.2
Total	72	12	22.4	16.7	9.6

SOURCE: U.S. Public Health Service.
neg. Negligible.

tons. Fuel oil burned in industry and for space heating also contains a small amount of sulfur—perhaps 0.25 per cent—which would result in about 1×10^6 tons of SO_2. (This may be somewhat underestimated.)

A figure for NO_X can be extrapolated from the coefficient used in Table 5, namely, 33.3 pounds per ton of contained carbon, or 1.37 per cent by weight. The industry and household category altogether consumed about 522×10^6 tons of contained carbon in fuel (1965), which would imply a production of 7×10^6 tons of NO_X. Information on carbon monoxides and on hydrocarbons unburned during combustion is hard to find, but one may probably assume coefficients for distillate oil in the industry-household sector similar to those for external combustion engines (Table 6), viz., 0.75 per cent for CO and 0.006 per cent for HC, by weight. Applying these coefficients indiscriminately to the total quantities of fuel burned, for which no other data are available, gives the numbers shown in Table 8, although this is a dangerous extrapolation. In particular, it may be too high for CO and too low for HC, at least where natural gas is the fuel.

Residuals from Materials Processing and Industrial Production

Describing materials flow and residuals production in manufacturing industries is particularly difficult. There are several reasons. First, there is a notable lack of direct information about quantities and qualities of industrial inputs, processes, and outputs, including residuals, which is both comprehensive and dependable. Second, industry generates an immense variety of residual products including most of those ultimately discarded by households, plus numerous others. Finally, while household, thermal power, and transportation activities are reasonably comparable across the country, the industrial mix of metropolitan areas varies drastically from place to place. There is also wide variation in a given industry even in the *same* place—as a function of age of plant, for example, where product output is held constant. For these reasons it is difficult to generalize meaningfully about industrial residuals (although some broadly applicable statements are possible). Thus at this point in time the following discussion is necessarily more illustrative than definitive. In many ways, materials balance concepts are more clearly applicable to industrial processes than to those in other economic sectors. But their comprehensive application must, like much other relevant analysis, await greatly improved data. At the end of this section we show, as an illustration, a detailed computed materials balance for one type of industrial plant under various conditions. This analysis yields some rather interesting conclusions.

For purposes of discussion it is convenient to subdivide the large and cumbersome "industrial" sector into several smaller subsectors, based on natural groupings among the materials involved in processing.

(1) One natural subcategory, belonging to the products of photosynthesis, comprises food processing, forest products, and organic chemicals (including petrochemicals and coal tar derivatives). The major classes of compounds involved are carbohydrates, fats and oils, proteins, and hydrocarbons, which are in turn made up primarily of carbon, hydrogen, oxygen, and nitrogen.

(2) A second large and slightly overlapping group comprises the inorganic chemicals and is based mainly on alkali metals (sodium, potassium, calcium, barium), halogens (fluorine, chlorine, bromine), boron, sulfur, phosphorus, plus the four basic constituents of organic chemicals.

(3) A third group comprises the ferrous metals (iron, nickel, and their alloys), aluminum, magnesium, copper, zinc, and lead.

(4) The last group includes the relatively inert materials, such as sand (quartz), gravel, stone, pumice, feldspar, clay, and gypsum. These are used primarily in building materials, insulation, abrasives, cement, refractories, ceramics, and glass.

Chart 3. Production and Disposal of Products of Photosynthesis

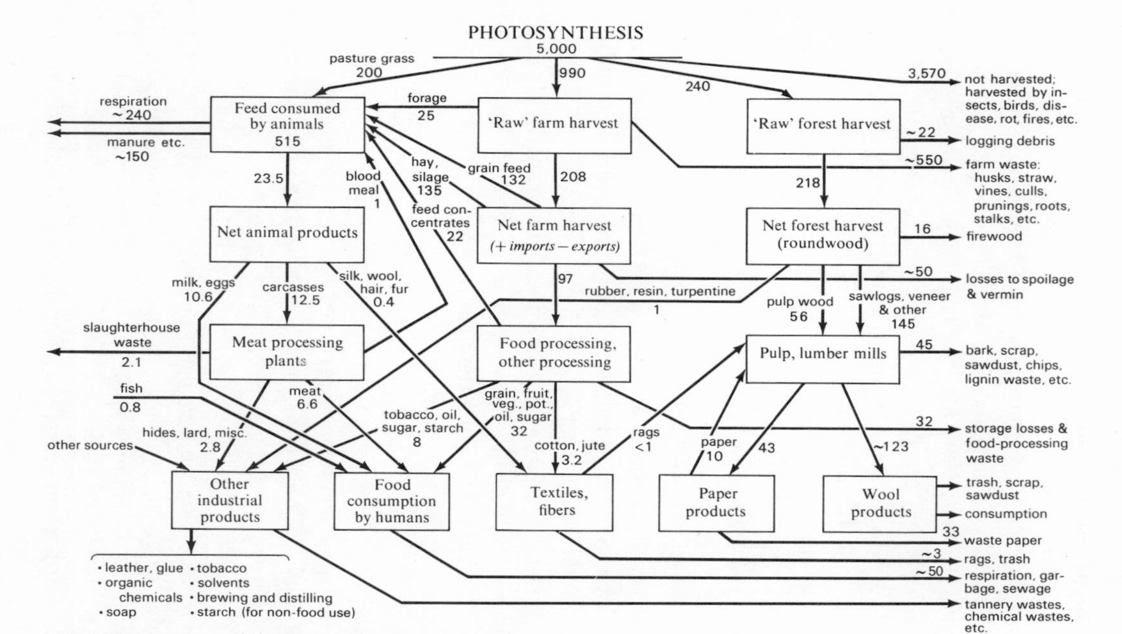

NOTES: This chart disregards fossil fuels, which are treated separately. All figures refer to millions of tons of dry organic matter.

Table 9. Materials Balance for Livestock

(Dry weigh x 10^6 tons)

Inputs of feed and forage	515.0
Output:	
Milk	9.6
Eggs	1.0
Meat and poultry (edible)[a]	6.6
Hides, lard, and other animal products	2.8
Wool, hair, and fur	0.4
Animal feed (recycled)	1.0
Bone meal and slaughterhouse waste	2.1
Manure and respiration losses (approximate)	~490.0

SOURCE: Estimates derived from U.S. Department of Agriculture statistics.

[a] "Edible portions" include some bone, skin, and fat which is later discarded in home preparation and appears as garbage.

The materials balance approach can be applied with particular effectiveness in the analysis of organic wastes from the processing of food and forest products. Chart 3 is a simplified "flow chart" for these activities. The total fresh weight of crops harvested for human consumption in the United States (excluding exports and industrial uses) is roughly 125×10^6 tons. This corresponds to about 59×10^6 tons of dry organic matter. An additional 515×10^6 tons of vegetable matter (dry weight) is harvested by and for livestock.[17] Wood and forest products harvested for consumption in the United States account for a further 218×10^6 tons (dry organic matter).

From the 515×10^6 tons (dry weight) of animal feed consumed annually in the United States, the net useful product consists of the items shown in Table 9.

Prior to the concentration of animal feeding activities in commercial feedlots and poultry farms to reduce labor, animal wastes (manure) were automatically recycled, thereby satisfying a major part of the nitrogen and phosphorus requirements of agriculture. Today, however, large quantities of manure are produced and accumulated where there is no market for the material. As a consequence, disposal has become an increasingly serious problem. Indeed the economies, which have hitherto justified feedlots, might be considerably diminished if the cost structure were broadened to include the disposal of the residuals by

[17] Including an estimated 200×10^6 tons of pasturage and 135×10^6 tons of hay and silage, plus 25×10^6 tons of forage.

returning them to the soil in some form, or (alternatively) to take into account the depreciation of the soil which occurs when the manure is not returned and the costs (external) imposed by runoff from feedlots into watercourses, especially after winter snowmelt.

Apart from the major (tonnage) residuals arising essentially from inefficiencies in the conversion of plant to animal calories, there are substantial residuals at the slaughtering stage. Although large meatpackers have stated that they utilize every part of the animal "but the squeal of the pig," this is an exaggeration. The edible portion of the animal is about 63 per cent for hogs, 59 per cent for beef cattle, and 48 per cent for sheep, or about 60 per cent overall. An additional 15 per cent consists of other useful or utilizable products such as hides, tallow, glue, and bloodmeal. The remainder, about 25 per cent of the total, is essentially a residual. Major meatpackers have achieved higher rates of economic recovery by incorporating some or most of this material in animal feeds and fertilizers; smaller local slaughterhouses cannot attain this degree of efficiency. Statistics in this area are difficult to find, but on a national basis probably at least 10 per cent of the live weight is not recovered and is discharged to the environment. The utilization of hides to manufacture leather products is also a major source of unrecycled residuals. Tannery wastes can be particularly obnoxious sources of water quality deterioration. The technology exists to reduce tannery wastes now discharged to watercourses by using organic solvents and enzymes to remove unwanted material, such as hair and fat, from the hides, thus producing a much more compact residue than is now the case. This is not done in general because the high capital investment required does not seem justified in a static or declining industry. Moreover, it would to some extent trade a water quality problem for an air quality problem.

The processing of food from plant origins for direct human consumption involves a smaller absolute quantity of organic waste. Table 10 shows the figures for harvested crops, excluding fibers, tobacco, crops destined for animals, and exports; it also shows the quantities of processed food available "for consumption" (prior to home preparation and cooking). The differences represent statistical inaccuracies, plus various processing losses, including spoilage due to vermin. In some cases, such as that of sugar beets, most of the weight of residuals is recovered (for cattle feed, for instance), but the residuals discharged to the environ-

Table 10. Materials Balance for Processing Foods of Vegetable Origin
(x 10^6 tons)

Food	Inputs, net crops harvested for human consumption (in U.S.)		Outputs: Food available for consumption		Outputs: Processing losses[a] and waste	
	Fresh weight	Dry weight	Fresh weight	Dry weight	Fresh weight	Dry weight
Grain	21.4	18.9	13.4	11.8	8.0	7.1
Potatoes and root crops	12.1	2.6	10.4	2.5	1.7	0.1
Beans and peanuts	2.2	2.0	1.6	1.4	0.6	0.6
Green vegetables	21.0	2.9	19.2	1.5	1.8	1.4
Fruits and nuts	16.7	2.4	14.1	1.8	2.6	0.6
Sugar (cane and beet)	49.3	27.7	10.2	10.2	39.1	17.5
Oil[b]	2.8	2.8	2.8	2.8	neg.	neg.
Total	125.5	59.3	71.7	32.0	53.8	27.3

SOURCE: *Agricultural Statistics*, 1966, Table 802.

neg. Negligible.

[a] Portions of these are recovered as by-products, but processing losses seem to be too low. For example, actual cannery plant data show residuals = 25–40 per cent of raw product into cannery. Unpublished study by Blair T. Bower.

[b] Vegetable oil is derived mainly from soybeans, cottonseed, and flaxseed; the pulp and meal are used as animal feed concentrates.

ment may be capable of exerting a high demand on assimilative capacity. Residual wastes from the processing of food products tend to be high in organics which reduce dissolved oxygen in rivers, for instance.

Although the demand for forest products, particularly pulp and paper, is rapidly increasing, the various processing stages are still comparatively inefficient on the average, resulting in enormous tonnages of unrecovered organic wastes. To begin with, only a small fraction of actual photosynthetic production is in fact harvested; the remainder is lost to disease, fires, insects, annual leaf fall, or is left to die and rot on the spot.

From "roundwood" to final wood and paper products, there are also substantial wastages en route. In paper-making, for example, which absorbed 56 $\times$ 10^6 tons of pulpwood in 1965 (dry organic matter), about 77 per cent of the input organic material is utilized as pulp; the remainder is residual. In a large modern integrated pulp and paper mill, some of this can be recovered and used as a raw material for organic chemicals—notably surfactants (detergents)—or as a growth medium for yeasts or bacteria which can, in turn, be used for animal feed. However, this degree of recovery is still relatively rare in the industry, and only about 15 per cent of the lignin residual liquors is recovered.[18] Accompanying the 1965 production of 33.3×10^6 tons of paper pulp (dry weight) was a residual bark, cellulose, and lignin liquor residual of 23×10^6 tons (dry organic matter), plus a large number of other chemicals used in the processing. This clearly puts a heavy demand on the assimilative capacity of watercourses, which are the major means of disposal at present.

The reuse of wastepaper recommends itself as a method of reducing the residuals problem arising from production of new pulp. Indeed, the production of paper products in 1965 was 43.7×10^6 tons, which means that about 25 per cent of the total production is recycled. However, as a means of reducing residuals discharge to the environment this approach has limitations, since much of the potential supply of wastepaper (e.g., newsprint) must be reprocessed to remove fillers, coatings, color, and ink. Whereas most original pulps produce an effluent load (excluding bark) of 20 to 200 pounds of suspended solids per ton of product—depending on the amount of bleaching required—de-inked wastepaper pulp yields 500 to 800 pounds of suspended solids per ton (25 to 40 per cent).[19] Thus, increased recycling of finished paper might not be very desirable, from the water quality standpoint, unless it were preceded by a substantial improvement in the technology for recovering waste lignin by-products. It would, however, contribute substantially to the alleviation of the solid residuals problem.

Sawmills, plywood plants, and veneering plants do not contribute large amounts of processing wastes, but there are substantial residuals in the form of bark, sawdust, chips, and scrap. Again, in large, efficient integrated plants most of this material can be utilized either in "pressed wood" or as chemical raw materials, or as inputs to pulp and paper manufacture. British Columbia now has a law that no new pulp mill

[18] See C. F. Gurnham, *Industrial Waste Water Control* (New York: Academic Press, 1965).
[19] *Ibid.*

will be permitted which uses raw wood for raw material. In the smaller local sawmills the residuals are disposed of (small proportions) or simply burned. At every stage in the utilization of wood products—notably in furniture manufacturing and construction—there is a substantial wastage, the amount tending to increase inversely to the size of the operation. Thus, residential and commercial construction is a major source of residuals; subsequent demolition also constributes substantial quantities. In areas of low population density, the burning of scrap lumber from construction poses no particular problem, but in dense conurbations the disposal of bulky demolition waste (a major constituent of the "trash" category of refuse) can be quite difficult.[20]

The chemical industry—both inorganic and organic—subsumes a large number of products which are used, or used up, in the preparation of other goods. A rather useful distinction can be made here between chemicals whose actual substance will ultimately become part of a final product, in some form, and chemicals serving other intermediate functions which do not appear physically in the final product.

The public is now well enough aware of the fact that agricultural chemicals such as pesticides, fungicides, and herbicides only serve their purpose (as presently used) when they are dispersed and/or degraded. Surface active agents such as detergents have also attracted a good deal of attention. It should be obvious, although it seldom seems to be pointed out, that the same is true of some dyes and pigments, virtually all industrial solvents and "carriers," softeners, fluxes, flotation agents, bleaches, cleaning agents, antifreezes, lubricants, explosives, and so forth. Tables 11 and 12 list some of the major tonnage organic and inorganic chemicals which are dissipated or dispersed, either immediately or gradually (as in the case of paints and persistent pesticides) into air, water, or soil in the course of *normal* use. We shall refer to these as "dissipative uses."

Apart from the existence of important dissipative uses in the chemical sector, there are also inevitable losses (residuals) in production and refining (see Table 13). This applies also, of course, to food processing and metallic ore reduction. However, as regards chemicals, it is noteworthy that there are often a number of intermediate stages *at each*

[20] The problem is aggravated by lumber which has been treated with petroleum products, i.e., creosote—piles, telephone poles, fence posts.

Table 11. Major Dissipative Uses of Industrial Organic Chemicals, 1963
(Approx. weight x 10^6 tons)

Solvents and thinners[a] (excluding gasoline and naphtha)	$\gtrsim$10
Antifreezes (ethylene glycol, alcohols, etc.)	~0.5
Pesticides and herbicides (DDT, 2,4–D, malathion, parathion, benzene hexachloride, etc.)	0.32
Surfactants[b]	1
Lubricating oil additives	0.2
Explosives (TNT, nitroglycerin, nitrocellulose, picric acid, ammonium nitrate, etc.)	?

[a] The tonnage of organic fluids used as solvents is very difficult to determine, since the major solvents (benzene, xylene, methyl-, ethyl-, or isopropyl alcohols, glycol ethers, acetone, methyl-ethyl-ketone, carbon disulfide, carbon tetrachloride, vinyl chloride, and various other chlorinated hydrocarbons) are also used for other purposes as well as intermediates. Total production of these chemicals including turpentine for *all* purposes (1963) was about 10 x 10^6 tons. In addition, an unspecified amount of gasoline is used as a solvent (e.g., for dry-cleaning).

[b] Surfactants are used extensively in the detergent and soap industries, along with fats and oils, sodium tripolyphosphate, sodium sulfate, and other inorganic chemical inputs. Most detergents and soaps are used in households, of course, rather than in industry. Total detergent production (1963) was just under 2 x 10^6 tons, and soap production was 0.5 x 10^6 tons.

Table 12. Major Dissipative Uses of Industrial Inorganic Chemicals, 1963
(Approx. weight x 10^6 tons)

Metal cleaning and pickling (H_2SO_4, HCl, etc.)	~1.25
Neutralization of excess acid, e.g., in paper manufacturing (CaO, $CaCO_3$, $NaCO_3$, NaOH)	$\gtrsim$1.5
Bleaches (H_2O_2, ClO_2, $CaClO_3$, ZnCl)	~0.1
"Antichlors" (sulfates, sulfites, NaOH)	>0.8
Water softening (CaO, etc.)	0.85
De-icing (roads and highways) (CaCl, NaCl)	~4.5
Paints and pigments (red lead [lead oxide], white lead [lead carbonate], TiO_2)	~0.8
Beet sugar refining (CaO)	0.6
Fluxing (fluorspar, cryolite, CaO)	~3.0
Insecticides and fungicides (lead arsenate, copper sulfate, etc.)	<0.1
Other (fabric finishing, tanning, photography, aerosol propellants, etc.)	?

one of which there may be significant generation of residuals. For instance, consider the chain leading to nylon, illustrated in Chart 4.

The cyclohexane route and the butadiene route involve the fewest intermediate stages (4), but a substantial fraction of all nylon is

Table 13. Typical Residuals as a Per Cent of Material Processed (by weight)

Industry	Aerosols	Gases and vapors	Typical residual percentage
Petroleum	Dust, mist	SO_2, H_2S, NH_3, CO, hydrocarbons, mercaptans	0.25 to 1.5
Chemical processes	Dust, mist, fume, and spray	Process-dependent	0.5 to 2
Pyrometallurgical and electrometallurgical processing	Dust, fume	SO_2, CO, fluorides, organics	0.5 to 2
Mineral processing	Dust, fume	Process-dependent	1 to 3
Food and feed processing	Dust, mist	Odorous materials	0.25 to 1

SOURCE: A. H. Rose, D. G. Stephan, and R. L. Stenburg, "Prevention and Control of Air Pollution by Process Changes or Equipment," in *Air Pollution* (Columbia University Press, 1961).

derived from phenol, which implies five or six intermediates. Nylon production in 1966 exceeded 0.5×10^6 tons.

A large number of other important chemical products are also derived via multistage processes, including synthetic rubbers, polyethylene, polypropylene, polyesters, PVC, polyurethane, insecticides, surfactants, etc. Needless to say, if 1 per cent of the material processed becomes a residual at each stage, the total residuals in producing a complex product may be 5 per cent or so of the final weight. Of course the rate varies, depending on a number of factors, notably the mode of handling and the volatility of the material. Thus, light petroleum and natural gas fractions are particularly subject to evaporative losses during refining and processing. If the overall rate to the atmosphere is only 1 per cent in each case (which may be low), the total quantities of hydrocarbons dissipated would be of the order of 8×10^6 tons, although it is hard to see how this figure could be very closely checked.

In the inorganic sphere, one of the largest tonnage operations is fertilizer production. We have not classed this as a dissipative use, although fertilizer consumption does involve dispersal, since presumably the active constituents constitute no hazard or pollution problem in the dispersed condition, except insofar as they are leached from the soil by surface water and thereby contribute to water pollution problems. Problems tend to arise, if at all, only during fertilizer processing,

Chart 4. Chemical Intermediates in Nylon Production

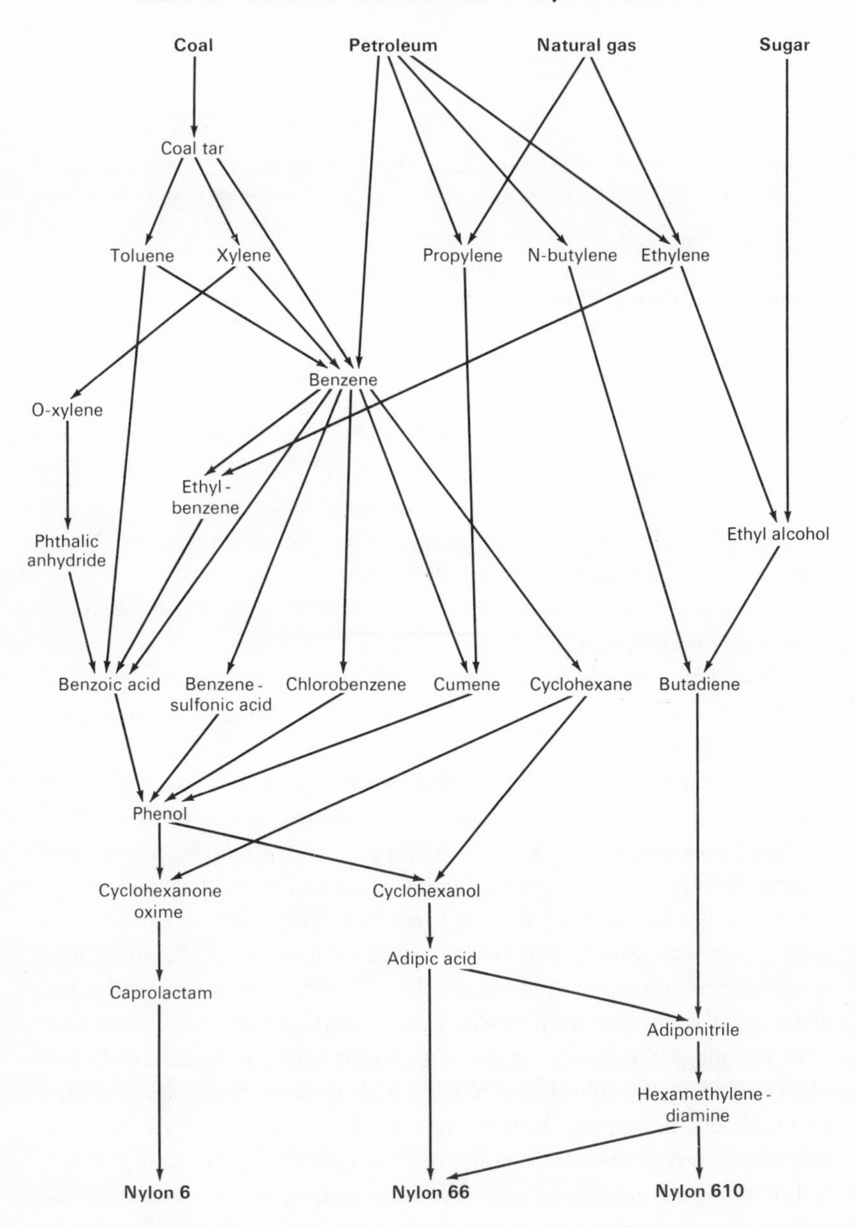

the most noteworthy example being the treatment of phosphate rock to produce commercial "superphosphate." This involves breaking down an insoluble complex material with the empirical formula

$$Ca_5\ (PO_4)_3\ (\text{halogen, OH})$$

where the halogen is usually fluorine, by treating it with sulfuric acid. The result is a mixture of $CaSO_4$ and $CaH_4(PO_4)_2$, with a small residue ($\sim$3 per cent) of fluorides or chlorides. If fluorides escape into the air they constitute a serious local air pollution problem, especially for commercial crops. This is currently a problem in Florida and Colorado.

Unlike the products of the chemical industries, the output of the mineral and metallurgical industries is almost entirely nondissipative. Residuals consist of bulky solids (e.g., slag), gaseous emissions associated with the energy conversion aspect of ore reduction—plus substantial quantities of particulates—and liquid wastes resulting from cleaning ("pickling") the metal during various phases of fabrication or treatment from ingot to final product, usually to remove oxide scales, which form when the hot metal comes in contact with air. Sulfuric acid is the major pickling agent for steel, but hydrochloric, nitric, and hydrofluoric acid are also used.

Slag, which once presented a serious disposal problem for steel mills, has in part become a valued by-product used mainly for road ballast and as aggregate for concrete products. It consists of a mixture of impurities from the ore plus fluxing agents such as limestone (used mainly for smelting iron), cryolite, or fluorspar.

Substantial quantities of particulates are produced by foundries, both as "flue dust"—some of which is recovered by Cottrell precipitators and wet scrubbers, because of its high metal content—and as soot from coal burning.[21] The latter is a major cause of sulfur dioxide and particulate emissions. The Public Health Service estimates that 2×10^6 tons of carbon monoxide, 9×10^6 tons of sulfur dioxide, 2×10^6 tons of NO_X, and 6×10^6 tons of particulates are produced by industry as a whole; presumably the majority of these residuals come from the energy-consuming ore reduction, refining, and other operations in the primary metals sector.

Waterborne residuals from the metallurgical industry are extremely

[21] Blast furnaces usually use coke, but other metallurgical operations such as reheating furnaces and rolling mills normally utilize coal.

difficult to recycle, because of the large bulk involved and the low price of the constituents. The spent pickle-liquor from steel mills consists primarily of ferrous sulfate ($FeSO_4$) plus some excess sulfuric acid. The latter can be neutralized with lime, but further treatment such as ponding or evaporation has not proved economically feasible to date. The quantity of such residuals can be deduced roughly from the amount of acid used for pickling (see Table 12). About 0.75×10^6 tons of H_2SO_4 were used in the iron and steel industry in 1963; this would correspond to roughly 1.1×10^6 tons of $FeSO_4$, and a loss of about 0.35×10^6 tons of metallic iron.[22] Similar processes are used in other metallurgical operations, particularly those of copper and brass mills. Other metallic sulfates discharged annually probably contribute an additional 0.75×10^6 tons, although the economics of recovery might be better in some cases. Apart from sulfate wastes, there are smaller but locally important quantities of other salts (chlorides, nitrates, and fluorides) and oily wastes from rolling operations.

Among the nonferrous metals (copper, lead, and zinc), ore beneficiation causes additional problems; roasting of sulfide ores creates very concentrated sulfur dioxide fumes, which are lethal to vegetation and have literally created sizable deserts around copper smelters in Tennessee and Montana. However, it has recently become economically feasible to recover most of this sulfur in the form of sulfuric acid, and many of these blights have already been substantially reduced (although the ecological consequences have by no means disappeared as yet).

Although the fourth group of materials processing industries far overshadows the others in tonnages of annual throughput, it is not a major source of environmental pollution problems. Such problems as do exist are mainly traceable to the energy conversion aspects of materials processing. For instance, 63×10^6 tons of portland cement were manufactured in 1963 from some 102×10^6 tons of mineral ingredients —mainly limestone, dolomite, quartz, and alumina, fused together at very high temperatures in a kiln by burning coke. The difference in weight is primarily due to the driving off of carbon dioxide by the heat. Plaster is similarly made from raw gypsum (hydrated calcium sulfate) by heating it to drive off the excess water; approximately

[22] In recent years there has been an increasing shift to hydrochloric acid, HCl, which is much more amenable to recovery and reuse.

11×10^6 tons of gypsum were thus converted in 1963. Bricks, tiles, refractories, ceramics, and glass also require intense heat for their preparation and thereby contribute to the demand for fossil fuels.

Potential Improvements through Treatment and Process Changes

There is a wide range of treatment methods available for application to waterborne industrial residuals. Some common types of treatment processes are as follows:

Screening*
Flocculation
Chemical coagulation*
Flotation
Sedimentation*
Centrifuging
Filtration
Stripping
Neutralization
Chemical oxidation*
Chemical reduction
Wet oxidation
Fermentation
Emulsion breaking
Evaporation
Distillation
Incineration*
Biological filtration*
Activated sludge*
Anaerobic digestion*
Stabilization lagoons*
Spray irrigation*

Processes to neutralize or remove gaseous or particulate emissions include:

Two-stage combustion
Afterburning
Catalytic oxidation
Recycling
Electrostatic precipitation
Wet scrubbing
Condensation
Absorption
Adsorption
Filtration

These processes change the chemical composition of the residuals or convert them from one form to another. Clearly none of them *eliminates* the residual.

The processes marked with an asterisk are also applicable to, and frequently used on, waterborne household residuals. Degradable organic wastes can be, and are sometimes, treated in a similar way, whether of industrial or household origin, although the treatment of industrial wastes can often be improved by combining them with household wastes for "fertilization" purposes.[23] Possible environmental problems associated with the residuals from these treatment processes are discussed in the section on household residuals.

Among the nondegradable industrial residuals—apart from combustion products—suspended and dissolved solids are the major ones. Suspended solids can be removed by sedimentation, with or without the aid of flocculants, such as polyelectrolytes, by filtration through various kinds of screens and filters, and by centrifuging. Dissolved solids can be removed, to any desired degree, by one or more of the processes of distillation, ion exchange, electrodialysis, and reverse osmosis. It is possible to obtain completely pure water from liquid residuals. However, it should be remembered that although "pure water" can be produced, there is still some sort of residual waste to be disposed of in some manner, either a concentrated brine or a semisolid sludge. The overall residuals management problem is not solved by the production of reusable water from these various processes. As we have frequently noted, ultimate disposal still remains a problem.

It should be emphasized that technological changes affecting residuals have not been, and are not now usually, instituted because of water problems or other environmental problems. In fact, almost all changes in production technology have thus far been stimulated by factors unrelated to environmental quality and have been developed without consideration for their external costs. But one can expect this situation to change. Various stimuli to management—such as effluent charges—in the instances where they have been used have resulted in process modification greatly reducing waste loads.[24] As more systematic means are developed to bring to bear on industry the external costs associated with the discharge of residuals into the environment, residuals generation and factors of control, including recycling of recovered materials and production of usable by-products, will receive more prominent consideration in process design. Studies of several industries have made clear that process design changes leading to recovery of residuals or their conversion to usable by-products can, in some instances, profoundly affect residuals generation.

[23] The reverse can also be true.

[24] For a detailed treatment of this experience see A. V. Kneese and B. T. Bower, *Managing Water Quality: Economics, Technology, Institutions* (Johns Hopkins Press for RFF, 1968).

In the following discussion we lean heavily upon the beet sugar industry as an illustration. This is not because it is the industry with the most important residuals problem, although as recently as 1950 it was estimated to discharge about 15 per cent of the organic wastes, measured in terms of "biochemical oxygen demand" (BOD), coming from all industries, but because its processes are comparatively simple, it has recently been intensively studied by Resources for the Future,[25] and we have been able to estimate a complete materials balance for representative plants using different processes.

The waterborne residuals load generated in pounds of BOD per ton of beets processed has been reduced greatly in the beet sugar industry as a whole in the last two decades by comparatively simple and economical alterations in processes. The main changes are the substitution of drying of beet pulp for storage of wet beet pulp in silos and the use of Steffens waste for the production of by-products. These changes reduce BOD generation by about 60 per cent. The other process change, i.e., a shift from cell type to continuous diffusers, is integrally related to recirculation of screen and press water. This further reduces the BOD generated by about 10 per cent.

Chart 5 indicates the main process and waste water residuals streams in representative beet sugar plants. Chart 6 shows residuals streams in a plant practicing no material or by-product recovery and discharging all of its liquid residuals to a watercourse. A few cases approaching this still exist. Chart 7 describes a plant in which all water is fully recirculated and there is no external discharge of waterborne residuals. There is one plant in the United States which uses basically this system; the others fall in intermediate positions. Charts 6 and 7 are helpful for understanding not only Table 14, which shows the estimated reduction in BOD discharge which took place between 1949 and 1962, but also the materials balance for a beet sugar plant presented later.

It should be noted that the "closed" plant requires treatment (in the form of clarification) for its recirculating water stream, despite the fact that materials recovery and by-product production have greatly reduced waterborne residuals. Even where opportunities to utilize process changes and increase recovery are favorable, some residual usually remains. The stream containing this residual may be treated,

[25] See G. O. G. Löf and A. V. Kneese, *The Economics of Water Utilization in the Beet Sugar Industry* (Resources for the Future, 1968). The materials balance described in later pages was calculated by George Löf.

Chart 5. Main Processes in a Beet Sugar Plant

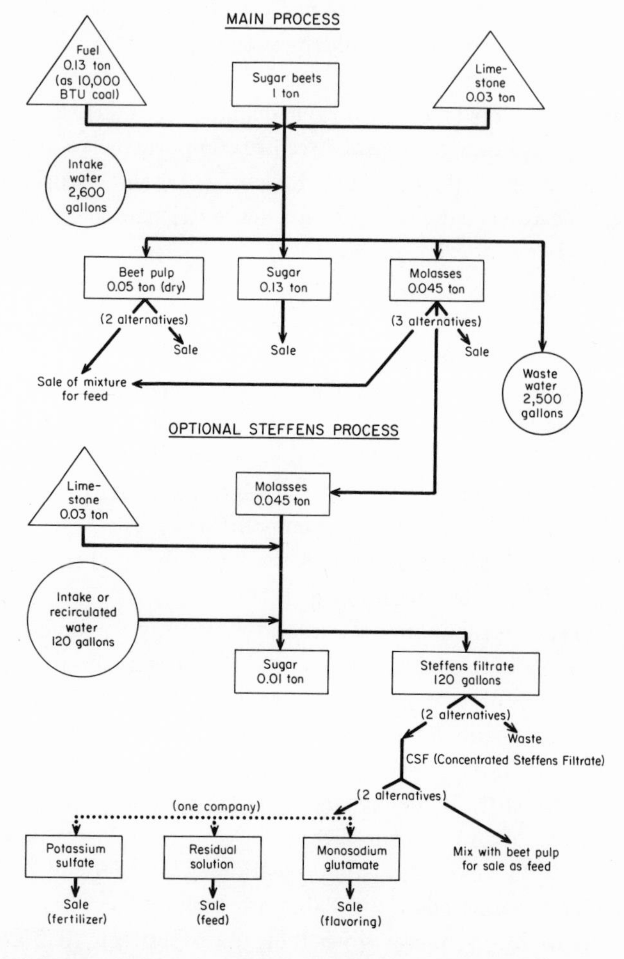

thus producing a solid or gaseous residual or changing the chemical composition of the waterborne residual.

To conclude this section we present a detailed materials balance calculation for two beet sugar production processes. One of these we term "high residual" and the other "low residual." Charts 8 and 9, showing the materials flow and residual materials, correspond to Charts 6 and 7, which show the water circulation streams. In a wet process industry like beet sugar production the two are closely related. Table 15 summarizes a few salient figures from the materials balance.

Chart 6. High Residual Beet Sugar Production Process

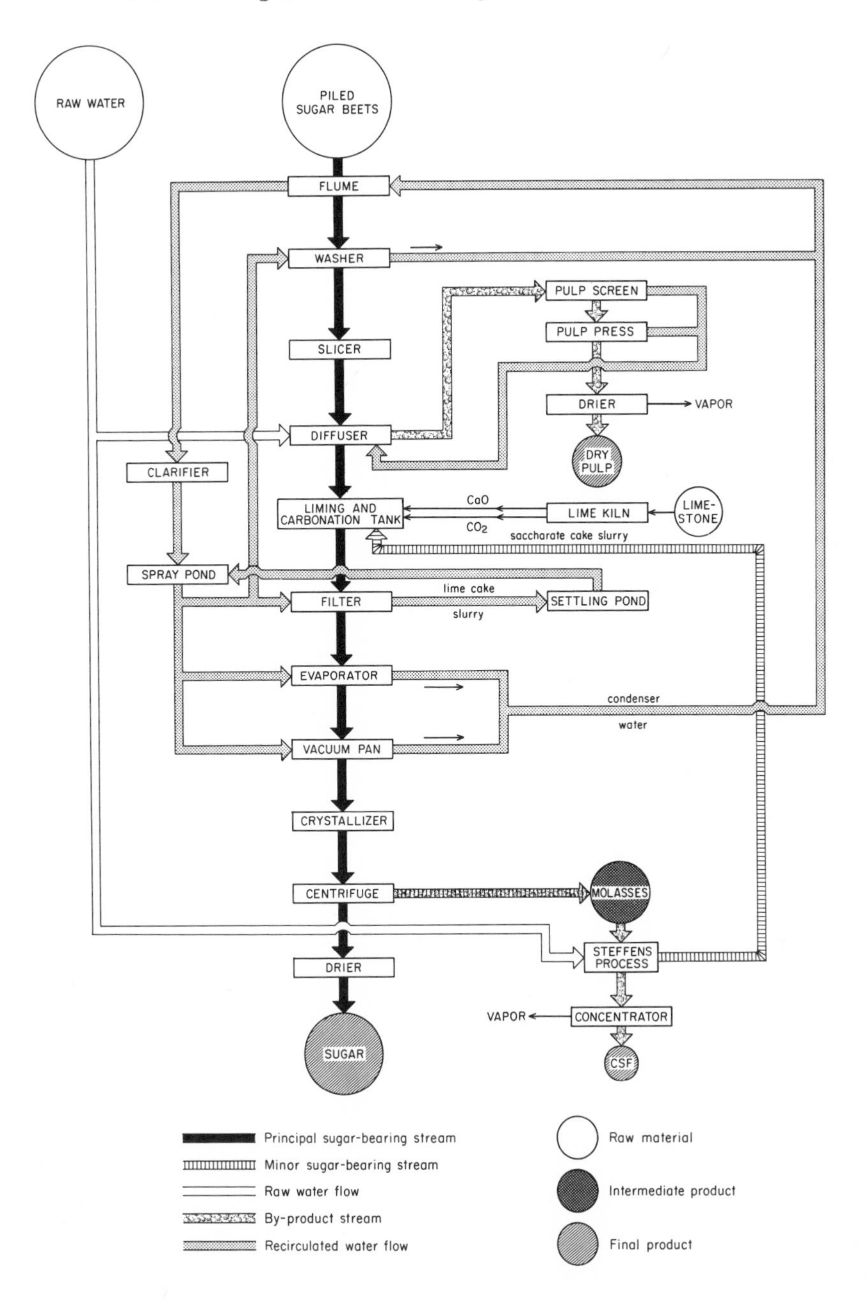

Chart 7. Low Residual Beet Sugar Production Process

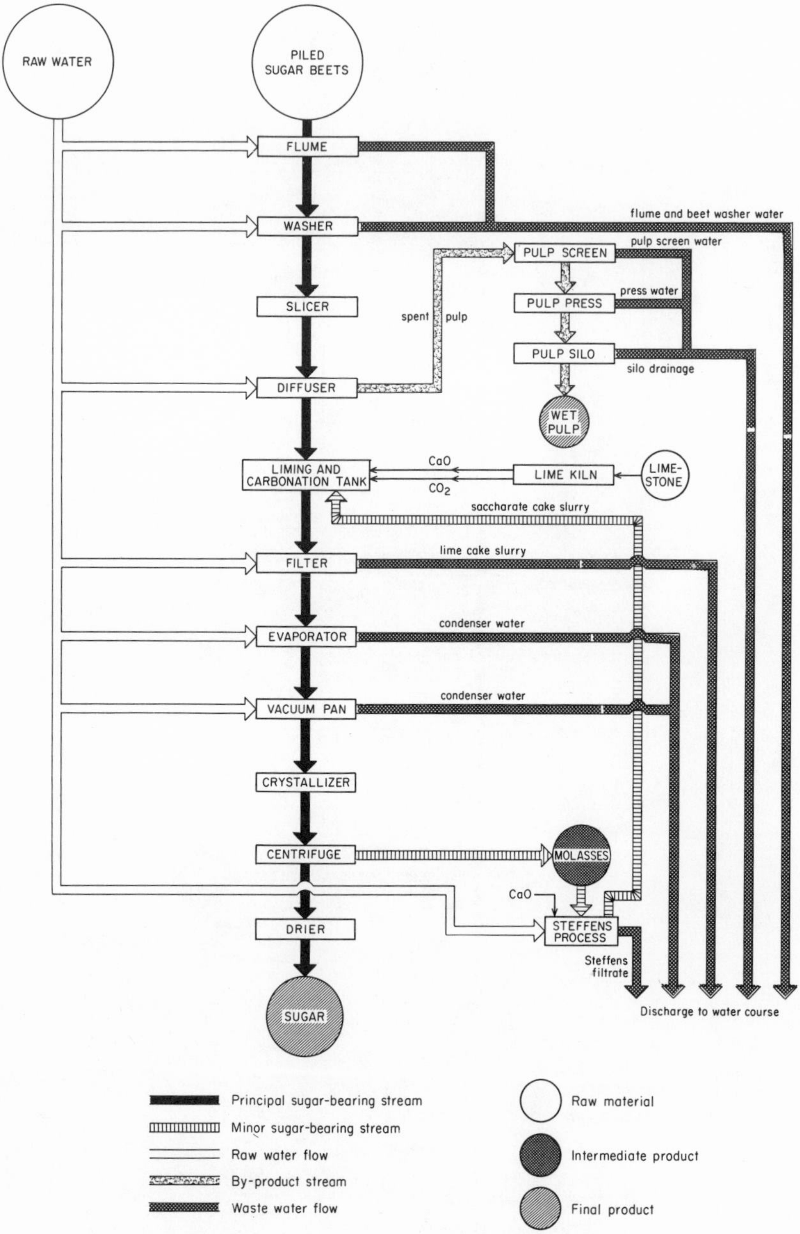

RAW WATER
PILED SUGAR BEETS
FLUME
WASHER
SLICER
DIFFUSER
LIMING AND CARBONATION TANK
FILTER
EVAPORATOR
VACUUM PAN
CRYSTALLIZER
CENTRIFUGE
DRIER
SUGAR
flume and beet washer water
PULP SCREEN
pulp screen water
PULP PRESS
press water
spent pulp
PULP SILO
silo drainage
WET PULP
CaO
CO_2
LIME KILN
LIME-STONE
saccharate cake slurry
lime cake slurry
condenser water
condenser water
MOLASSES
CaO
STEFFENS PROCESS
Steffens filtrate
Discharge to water course

Principal sugar-bearing stream
Minor sugar-bearing stream
Raw water flow
By-product stream
Waste water flow
Raw material
Intermediate product
Final product

Table 14. Estimated Reduction of BOD in Beet Sugar Processing, 1949 and 1962

(1,000 pounds per day)

Type of waste	1949					1962				
	BOD generated[a]	BOD removed by process changes[b]	BOD removed by waste treatment	Total BOD removal	BOD discharged	BOD generated[a]	BOD removed by process changes[b]	BOD removed by waste treatment	Total BOD removal	BOD discharged
Flume and washer water	510	n.a.	n.a.	100	410	710	n.a.	n.a.	270	440
Cooling water and condensate	80	n.a.	n.a.	10[c]	70[c]	110	n.a.	n.a.	30	80
Pulp screen and press water	550	50	70	120	430	840	630	60	690	150
Silo drainage	1,390[d]	660[e]	140	800	590	1,940[d]	1,920[e]	10	1,930	10
Lime cake slurry	730	0	350	350	380	1,030	0	960	960	70
Steffens filtrate	610	160[f]	80	240	370	770	560[f]	160	720	50
Total BOD	3,870			1,620	2,250	5,400			4,600	800

Source: G. O. G. Löf and A. V. Kneese, *The Economics of Water Utilization in the Beet Sugar Industry* (Resources for the Future, 1968).

Note: Based on 158,000 tons of beets per day processed by 58 plants operating in 1962. In 1949, 113,000 tons per day were processed. To enable direct comparison, the data for 1949 were extrapolated to production of 158,000 tons per day, assuming constant proportions.

n.a. Not available.

[a] Based on BOD per ton of beets sliced in an "unimproved" plant, from "Industrial Waste Guide to the Beet Industry," U.S. Public Health Service, Dec. 1950.

[b] By process changes and recirculation.

[c] Based on estimated 10 per cent reuse as diffuser make-up water.

[d] BOD which would be generated if all spent pulp were handled in silos, i.e., no pulp drying.

[e] BOD not generated because of use of pulp driers.

[f] By recycle-to-production process and CSF production.

Chart 8. High Residual Beet Sugar Production Process—No Recirculation

(In pounds per ton of beets processed except where otherwise stated) Intake: 5250 gallons/ton sliced beets—regular; 175 gallons/ton sliced beets—Steffens additional.

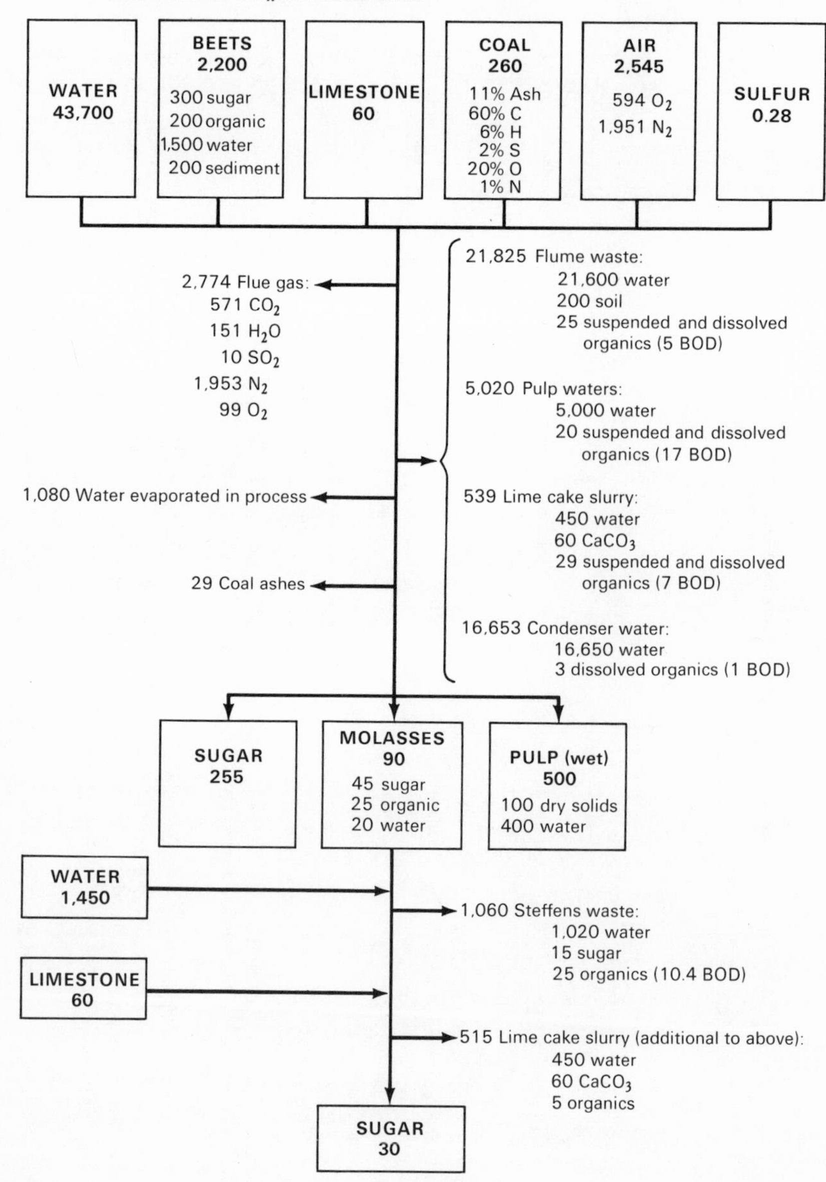

It will be noted from the table that from the point of view of waste residuals a large reduction in organic residuals was purchased at the expense of a comparatively small increase in potentially harmful gas and inert solids. Again the interdependency between the residual waste streams is revealed. Considering the environment in which most beet sugar factories operate, away from large cities but near small streams with very limited capacity to assimilate organic wastes, the tradeoff shown is probably favorable.[26] A conventional study which looks only at air, water, or solids problems individually would never reveal such tradeoffs, nor would it permit their examination from the point of view of the full range of industrial costs, internal and external.

Residuals Associated With Final "Consumption": Households

The household is where final consumer goods are utilized and give rise to the residuals remaining at the end of the materials flow through the economy to the consumer. Of course, all material inputs to the consumption sector except those added to inventory or recycled back into the production system become residuals which must be returned to an environmental medium in one form or another. We have already considered one important class of household activities giving rise to large amounts of residuals—the transportation of members of the household. We included this discussion in the energy conversion section because the interdependencies between the various types of residuals and possible methods for residuals control can be better analyzed in the context of overall urban transportation systems.

The other major residuals from household activities are gases resulting from space heating and home incineration or other combustion, sewage (solids in water suspension but containing also dissolved solids like chlorides and nitrates), and solid wastes including garbage and mixed refuse as well as junk autos and the like.

Generally speaking, the space heating and incineration[27] activities of

[26] Combustion of coal in beet sugar plants is conducted on too small a scale to make recovery of sulfur practical. (See section on thermal power.) However, fuel substitution might be feasible in certain instances.

[27] We do not include here the residuals from incinerators after collection of solid wastes. These are discussed below.

Chart 9. Low Residual Beet Sugar Production Process with Extensive Recycle

(In pounds per ton of beets processed except where otherwise stated)
Intake: 270 gallons/ton sliced beets—regular; 128 gallons/ton sliced beets—Steffens.

Coal quantity based on 260 for simple plant + 60 for pulp drying + 25 for CSF production (evaporation). Coal assumed 10,000 Btu/lb., 11% ash, 60% C, 6% H, 20% O, 2% S, 1% N. Coal assumed to provide all the plant heat requirements, including pulp dryer.

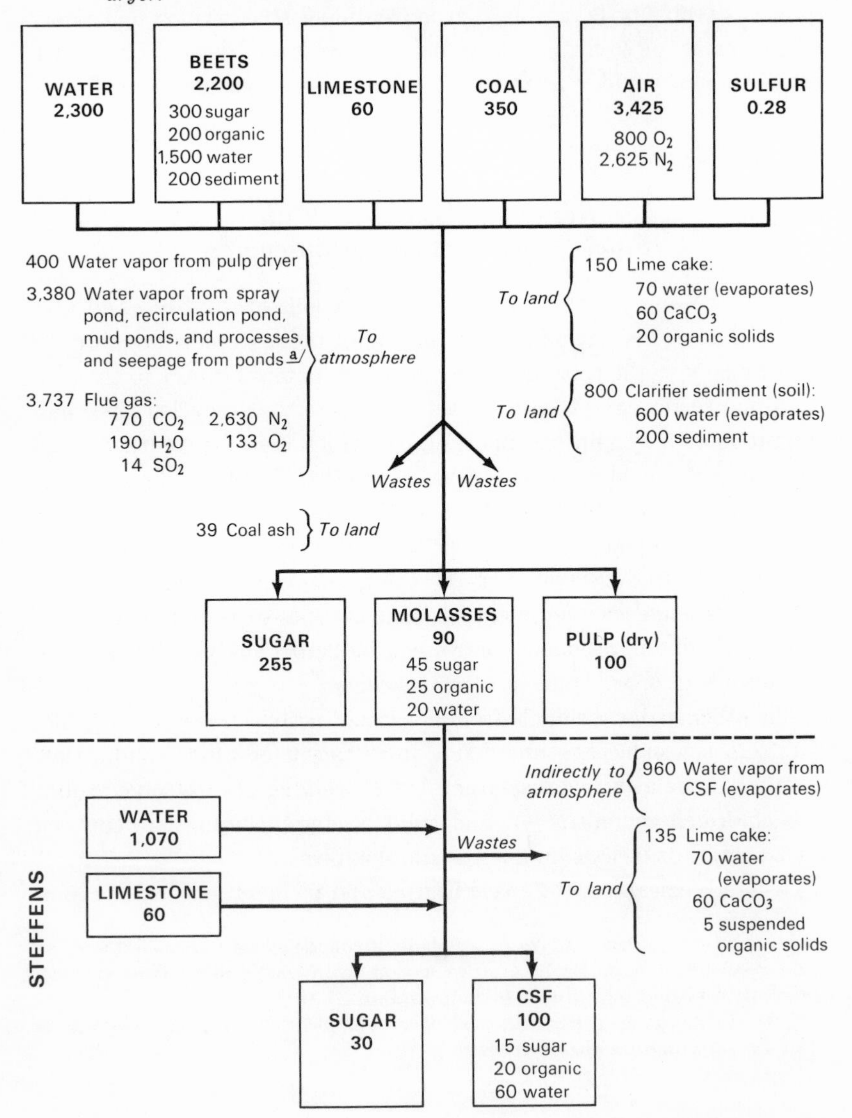

Table 15. Selected Figures from Materials Balance for Two Beet Sugar Processes

(In pounds per ton of beets processed)

Inputs and outputs	High residual process	Low residual process
Raw material inputs:		
Beets	2,200	2,200
Limestone	60	60
Coal	260	350
Sulfur	0.28	0.28
Product outputs:		
Sugar	285	285
Pulp	100[a]	100[b]
Concentrated Steffens filtrate (used for stock feed to recover monosodium glutamate and potassium sulfate)	0	100
Waste residuals:		
SO_2	10	14
$CaCO_3$	120	120
Coal ashes	29	39
Organics	122	25
Soil	200	200

[a] Dry weight of wet pulp.

[b] Dry.

households do not appear to contribute a major portion of the gaseous residuals found in the atmosphere of cities, nor are gases generally a very significant portion of the weight of residual materials stemming directly from households. However, there is considerable variation among cities in this regard. In Table 16, dealing with Sacramento, which reflects conditions perhaps characteristic of areas where there is little heavy industry, relatively clean fuels are used, and home incineration is relatively uncommon, the preeminent position of autos in regard to gaseous emissions is clearly evident. Also it is apparent that residential incineration and fuel combustion contribute only marginally

Notes to Chart 9

Extra fuel for Steffens process included in original total quantity shown.

Lime cake from Steffens process actually combined with lime cake from straight process. Quantities shown separately.

[a] The 70 lb. of water in the lime cake and 600 water in the clarifier mud evaporate from ponds, and *are included* in the 3,380 total evaporation to atmosphere.

Table 16. Gaseous Emissions in Sacramento County, 1964

(Tons per day)

Source	Contaminant			
	Carbon monoxide (CO)	Hydro-carbons (HC)	Sulfur dioxide (SO_2)	Oxides of nitrogen (NO_X)
Stationary source:				
Downtown business	2	18	1	4
RR station and airport	1	3	1	3
Travis AFB	7	5	<0.5	1
Wood burning	3	<0.5	<0.5	<0.5
Asphalt plants	2	4	1	4
Aerojet-General Corp.	2	1	<0.5	1
Residential and incin.	2	3	0.3	2
Municipal refuse	1	1	<0.5	<0.5
Motor vehicles	760	140	2	45
Total	780	175	6	60

SOURCE: Aerojet-General Corporation, *California Waste Management Study,* a report to the State of California, Department of Public Health, Report No. 3056 (Final), Azusa, Calif., Aug. 1965.

to total emissions. However, the type of fuel used in homes is very significant in this respect. Natural gas is a relatively "clean" fuel in the sense that combustion is rather complete. For example, the relative quantities of particulates emitted to the atmosphere per Btu of fuel burned are 1 for coal, 0.6 for oil, and 0.15 for natural gas. Also, natural gas is relatively free of sulfur compounds while certain coals and residual fuel oils contain up to 3 or 4 per cent of sulfur, as we have already seen. Consequently, in cities where space heating with lower quality fuels is dominant, the contribution of households to overall emissions may be considerably greater. Table 17 shows, for example, estimates of emissions from various sources in New York City. In general it is difficult and expensive to control harmful emissions from households by treatment. Fuel substitution or centralized provision of heating services—via central steam plants or electricity—would usually be more practical.

Another point illustrated by Table 17 is the importance of refuse incineration as a source of emissions, especially particulates and hydrocarbons. Some of these come from central incinerators, but in New

Table 17. Gaseous Emissions in New York City, 1964
(Tons per day)

Source	Gaseous residual (emission)				
	Carbon monoxide (CO)	Hydro-carbons (HC)	Sulfur dioxide (SO_2)	Oxides of nitrogen (NO_X)	Particu-lates
Stationary source:					
Electric power	1	4	754	254	35
Industrial	2	2	140	54	26
Commercial, inst. and large apts.	76	21	678	212	37
Small residential	4	4	67	78	24
Refuse combustion	291	120	6	7	75
Misc.[a]	750	31	n.a.	705	7
Motor vehicles	3,784	695	14	162	22
Total	4,908	877	1,659	1,472	226

SOURCE: Tri-State Transportation Commission, "Electric Power and Fuel Consumption, 1965–1985," New York, July 1967.

n.a. Not available.

[a] Mainly gasoline marketing, commercial dry-cleaning, etc.

York and other large cities apartment house incinerators are also a significant contributor. Combustion processes in these incinerators are usually inefficient and particulates are poorly controlled, if at all.

In the control of gaseous residuals from household activities, there are important tradeoffs with other residual discharges, as to both type and spatial distribution. We will return to this point later.

One of the major components of household residuals is sewage. Most of the dry weight of sewage (about 0.55 lb. per person per day) from households not using garbage grinders is composed of human excrement. In addition there are various organic and inorganic substances which result from cooking and washing operations (see Table 18). From households using garbage grinders, the total dry weight of sewage is about doubled. Actually the materials balance implies that the total dry weight of all organic wastes (sewage plus garbage) must be equal to the dry weight of food products entering the household, less the carbon content of CO_2 produced by respiration and minor corrections for

Table 18. Average Per Capita Solids and BOD_5 in Domestic Sewage
(Grams per capita per day[a])

State of solids	Mineral	Organic[b]	Total	BOD_5
Suspended:	25	65	90	42
a. Settleable	15	39	54	19
b. Nonsettleable	10	26	36	23
Dissolved	80	80	160	12
Total	105	145	250	54

Source: G. M. Fair and J. C. Geyer, *Water Supply and Waste Disposal* (John Wiley & Sons, 1956), p. 563.

[a] 1 gram per capita = 2.2 lbs. per 1,000 population.

[b] Of the organic matter in average domestic sewage about 40% is composed of nitrogeneous substances, 50% of carbohydrates, and 10% of fats. The detailed composition (carbon, hydrogen, oxygen, nitrogen) after primary treatment varies, depending on the process used: carbon content ranges from 52% to 67%, hydrogen from 7% to 9%, nitrogen from 3% to 4%, and oxygen from 21% to 38%. See R. Rickles, "Pollution Control 1965," Chemical Process Monograph No. 10, Noyes Development Corp., 1965.

annual accumulation (increase in population) and deaths (see Table 19).

Of most direct concern in connection with household wastes are the degradable organic materials which impose a demand on the dissolved oxygen of waters to which they are discharged. This is usually measured in terms of the 5-day biochemical oxygen demand, BOD_5. Since the BOD_5 varies directly with temperature it is usually measured on a common temperature base of 20° centigrade. It is convenient to think of BOD_5 as a substance in the water—a measure of the degree of residual which can be added or removed. Of increasing significance too are the plant nutrients which are the final stage of the breakdown of degradable waste. Added to this plant nutrient contribution from human excrement are the phosphate builders used in detergents.

Numerous detailed studies of the techniques used in treating domestic sewage are available, so we will not discuss them extensively. The most basic processes, invented more than 50 years ago, use the principles of settling solids, sometimes with the assistance of chemical flocculents and biological degradation. With these methods about 90 per cent of BOD_5 can be removed. Higher rates of removal can be achieved

Table 19. Hypothetical Materials Balance for Humans, 1963
(Dry weight x 10^6 tons)

	Carbon	Oxygen	Hydrogen	Nitrogen	Total
Food inputs:					
Carbohydrate	14.74	16.39	1.99	0	33.12
Fat	7.36	1.11	1.11	0	9.63
Protein	3.92	1.67	0.51	1.16	7.25
Total	26.01	19.17	3.65	1.16	50.00
Outputs:					
Garbage	~5.99	~3.12	~0.80	~0.09	10.03
Respiration[a]	~12.84	~9.79	~1.84	0	24.46
Sewage[b]	~6.44	~6.06	~0.91	0.98 + 0.06	14.39
Losses due to death[c]	0.46	0.12	0.06	0.06	0.70
Added biomass (population growth)	0.28	0.07	0.03	0.03	0.43

Note: All figures prefaced by (~) are estimates based on plausible allocations of protein, carbohydrate, and fat.

[a] Not including oxygen from the air; proportions based on combustion of sugar ($C_6H_{10}O_5$) yielding CO_2 and H_2O.

[b] Sewage solids estimated at 0.55 lb. per capita per day. See Table 18 for breakdown into components.

[c] Assuming the population increase is 1.2% (of the biomass) per year, and the death rate is 2% (of the biomass).

by chemical and filtration processes. Some of these (like adsorption with activated carbon) can achieve up to 98 or 99 per cent removal of organic materials from domestic sewage. Costs vary approximately in proportion to $\frac{1}{1 - E}$ where E is the removal efficiency in percentage.[28]

A kind of ultimate treatment would involve distillation of the final effluent, producing pure H_2O as one final product. But even this does not mean that the residual materials are thereby eliminated; a solid or semisolid sludge remains which must either be used to produce reusable materials or be dispersed into the environment. The most common pro-

[28] For a detailed discussion of advanced treatment techniques together with cost and performance estimates, see A. V. Kneese and R. J. Frankel, "The Economics of Water Reclamation," in *Conservation and Reclamation of Water*, proceedings of a Symposium sponsored by The Institute of Water Pollution Control, held on Nov. 29, 1967, in London.

cedure is to "digest" sludges resulting from sewage treatment in heated anaerobic tanks which produce CO_2 and methane gas (the latter being used for fuel in the plant or burned off), a supernatant liquor which contains most of the nitrogen and phosphorus originally contained in the sewage, and a stable organic material (mostly cellulose) which is referred to as the "digested sludge." The last is most often used as landfill or simply dumped. The supernatant liquor is usually discharged to a convenient receiving watercourse, where the additional nutrients may cause excessive fertilization and unpleasant or harmful algae blooms.

Another method for disposing of sludge involves its combustion (some techniques like the Zimmerman process do not require its prior digestion). Combustion—whether for power or not—results in the discharge of gases and still leaves a relatively small residual amount of solids which must be disposed of. If the sludge has not been digested prior to combustion, the solid residual will be rich in plant nutrients[29] and therefore potentially useful as fertilizer. Again we see the strong interdependencies between the control of liquid, gaseous, and solid waste streams and the important role which recycling may play in controlling external cost. For example, while it might seldom, if ever, be economically justified, distillation of municipal sewage, with the solids resulting from perfect combustion reused as fertilizers, would result in only CO_2 and H_2O residuals being discharged into the environment. The spectrum of sewage treatment methods currently employed in the United States is summarized in Chart 10.

In addition to gaseous residuals and sewage, households produce

[29] In some areas where water pollution is reasonably effectively managed, sludge disposal is deemed the major unsolved problem. The Ruhr area of Germany and the Metropolitan Sanitary District of Greater Chicago are examples. In Chicago, until 1961, much of the wet undigested sludge (about 900 tons a day—dry weight —produced at the West-Southwest sewage treatment plant) was deposited into huge sludge lagoons along the main canal, where the discharge of gases caused a great odor nuisance. Some sludge is still air-dried and continues to cause significant odor problems. Also about 500 tons a day are heat-dried and sold as fertilizer at a net loss to the district of $38 per ton. The district is also experimenting with a 200-ton-a-day Zimmerman plant. The ash resulting from this process is presently dumped. See statement of Vinton W. Bacon, general superintendent, Metropolitan Sanitary District of Greater Chicago, to Natural Resources and Power Subcommittee, Committee on Government Operations, House of Representatives, Congress of the United States, at Chicago, Ill., Sept. 6, 1963.

Chart 10. Sewage Treatment in the United States (as per cent of population)

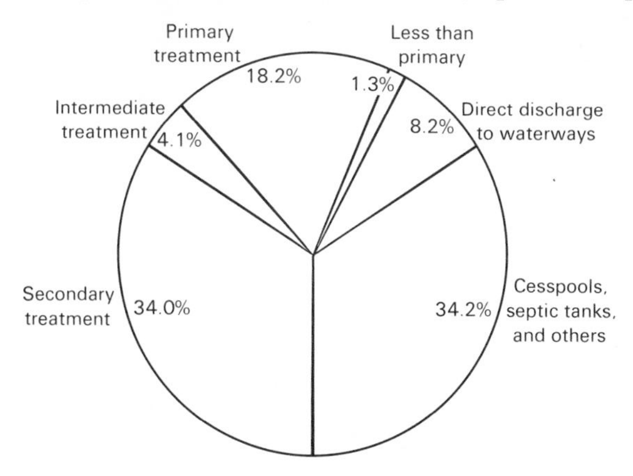

residuals which are in solid form, including garbage, rubbish, and ashes. These terms are defined as follows:

(1) Garbage: Residuals from preparation, cooking, and serving of food.

(2) Rubbish—

Combustible: Paper, plastics, cartons, boxes, barrels, wood, excelsior, tree branches, yard trimmings, leaves, wood furniture, bedding, dunnage, etc.

Noncombustible: Aluminum foil, tin cans, metal furniture, dirt, glass, crockery, etc.

(3) Ashes: Incombustible residue from fires used for heating and on-site incineration.

Altogether, the average member of a household threw away in 1965 somewhere between 3 and 5 pounds of solid residuals per day. See Table 20 and Chart 11.[30]

These residuals are disposed of in a variety of ways in urban areas. Backyard burning and burial have greatly diminished and the pattern has shifted historically to use of neighborhood dumps and then central incinerators and countryside landfills. Use of overall land disposal sites

[30] Substantial, but unknown, amounts of urban solid wastes are not collected. So the figures shown include significantly less than the total solid wastes generated. Improved data on solid wastes are badly needed.

Table 20. Total Mixed Refuse Collected in Five U.S. Cities, 1957–58

City	Pounds per capita per day
Los Angeles, Calif.	4.61
Washington, D.C.	4.36
Seattle, Wash.	3.75
New York, N.Y.	3.63
San Francisco, Calif.	2.18

SOURCE: Aerojet-General Corp., *California Waste Management Study.*

is by far the most common method. But the present practices still give rise to major external costs: indiscriminate dump sites still litter the countryside; odor, smoke, insects, and rodents accompany open dumps (which are often burning to reduce volume and help control insects and rodents); contributions to air pollution arise from improper incineration and uncontrolled burning; and water quality degradation results from drainage through dumps and direct disposal of solid wastes into waterways. As part of a study of the solid waste problem by the Maryland Department of Health a survey was made in July 1966 of solid waste disposal practice in all 23 Maryland counties and the city of Baltimore. Practices in the State could perhaps be considered typical of those in the megalopolitan East. At least 40 per cent of the 155 disposal sites were clearly producing external costs.[31] And this is probably very good performance in relation to the national average. Only a very small percentage of land disposal sites are "sanitary land fills" in the sanitary engineer sense of the term.

The costs of solid residuals collection and disposal, as presently practiced, are high. It is estimated that annual local government costs for collection and disposal are $1.5 billion. This is exceeded in local government budgets only by expenditures for schools and roads. In addition it has been estimated that the annual expenditures of the private sanitation industry are nearly as high.[32] Of the total cost of disposal about 80 per cent is for collection. Collection is labor-intensive and technologically primitive.

Collection costs are also highly sensitive to certain variables. A cross-

[31] See *Collection and Disposal of Solid Wastes: A Maryland Program,* Maryland State Department of Health, Aug. 1, 1966.

[32] W. E. Gilbertson and R. Black, *Solid Wastes,* National Commission on Community Health Services, 1964.

Chart 11. Projection of Refuse Production Trends

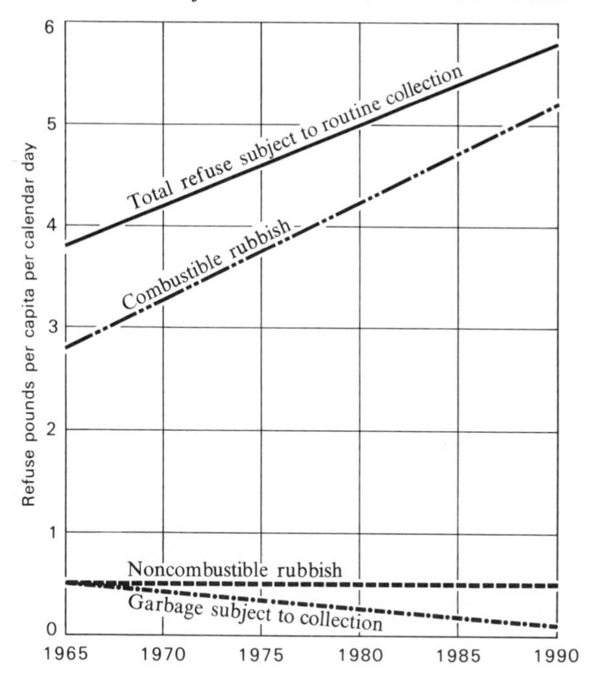

section statistical analysis of St. Louis found that an increase from two weekly pickups to three increased collection costs by nearly one-third, while moving pickup location from the curb to the rear of the house almost doubled the cost.[33]

Because collection is such an expensive and cumbersome process, it is attractive to think in terms of different and less labor-intensive collection systems. One proposal that recurs frequently in the literature is to grind solid wastes and deliver them to the sewers. Since sanitary sewage is now over 99 per cent water, the existing sewers could accomodate a considerable increase in solids without an expansion in capacity. Additional grinding could be done in the home or at grinding stations located relatively close to the points of pickup. In the latter cases various organic substances in addition to garbage could be ground up. This would reduce (but certainly not eliminate) collection costs, and it would require additional sorting.

This procedure would, of course, increase the waterborne residuals

[33] W. Z. Hirsch, "Cost Functions of an Urban Government Service: Refuse Collection," *Review of Economics and Statistics*, Vol. 47, No. 1 (Feb. 1965).

and put an additional loading on the sewage treatment plants and/or the stream. The interdependency of various residuals streams and means of control is again evident. On the other hand, it would reduce the load on the transportation system by reducing the number of collection trucks and thus increasing the overall speed of surface traffic.[34] This in turn would tend to reduce gaseous emissions somewhat. Savings realized at the collection end might possibly compensate for increased treatment costs.

As far as disposal is concerned, combustion looks favorable because, as shown in Chart 11, so large a portion (roughly 80 per cent) of the total rubbish and garbage is combustible. Incineration reduces the total volume of the solid wastes to about one-fifth of that before combustion.[35] Consequently, from a narrow cost minimization point of view it would be desirable to conduct the incineration very close to the source (in apartment house incinerators and backyards). This, however, results in converting the major part of the residuals to gases and airborne particulates, which in highly developed areas are likely to give rise to large external costs. If efficient techniques for controlling emissions from small incinerators could be worked out, burning close to the source would be very attractive.

In recent years solid residuals from households have been substantially increased by a heavy trend toward nonreturnable containers of all types. For example, the use of one-way bottles almost doubled between 1965 and 1966.[36] In many instances, also, nonreturnable containers are being made of less degradable materials than previously—primarily plastic and aluminum. Overall, in 1966, 48 billion cans, 26 billion bottles and jars, and 26 billion metal and plastic caps were produced, virtually all of which were eventually discarded. Rates of reclamation and reuse are rather low. About 10 per cent of the plastic produced is recovered and about 15 per cent of the rubber products

[34] Experiments at the University of Pennsylvania indicate that ground-up solid residuals (including metal and glass) can be successfully pumped as a liquid slurry through surprisingly small diameter pipes. A 2-inch pipe would suffice for a city of 10,000 or 15,000. This might provide an alternative to conventional collection and transportation but would present a problem of liquid disposal.

[35] At several locations in Europe incineration is carried out in thermal electric plants. At Munich's Nord power station, for example, refuse is burned along with powdered coal. Electrostatic precipitators are used to control particulate emissions in the stack gases. The electric power system charges the refuse collection agency $1.50 per ton to burn the refuse delivered to the plant.

[36] *Chemical Week*, Vol. 99, No. 19 (1966), p. 44.

(mostly automobile and truck tires).[37] Of the 43×10^6 tons of paper products produced each year, less than one-third is reclaimed.

A major type of solid residual, which poses quite special problems, is worn-out and discarded automobiles. In the middle sixties there were about 90×10^6 automobiles and trucks operating, with annual production of about 9×10^6 vehicles, 6 to 6.5×10^6 vehicles scrapped each year, and about 2.7×10^6 vehicles added to inventory.[38] This accounted for about 30 per cent of production.

Inputs to the "automotive transportation" subsector accounted for—

1.6×10^6 tons of new rubber, of which about 1×10^6 tons were tires for new motor vehicles and the remainder were replacements or rubber recaps (60 per cent of total U.S. rubber consumption).[39]

17×10^6 tons of steel for automobiles and trucks (15.5 per cent of U.S. consumption).[40]

1×10^6 tons of fabrics, carpets, paint, insulating materials, plastics, and other nonmetals.[41]

0.5×10^6 tons of glass.[42]

0.6×10^6 tons of lead for batteries (30 per cent of U.S. consumption).[43]

0.15×10^6 tons of copper (9 per cent of consumption).[44]

0.5×10^6 tons of aluminum, zinc, nickel, and other metals.

[37] "Environmental Pollution: A Challenge to Science and Technology," report of Subcommittee on Science Research and Development, Committee on Science and Astronautics, U.S. House of Representatives, 89th Cong., 2d sess. (U.S. Government Printing Office, 1966), p. 44.

[38] Based on the anticipated growth in total vehicle population from 1960 to 1970. Estimates published by the Federal Power Commission, Feb. 1967, based on various projections by the Bureau of Public Roads, the Bureau of the Census, RFF, and the American Automobile Association.

[39] Data on rubber, glass, and other nonmetals were inferred from a variety of sources. Breakdown on weight composition of automobiles from G. A. Hoffman, *Automobiles, Today and Tomorrow,* The Rand Corp., RM–2922–FF, Nov. 1962. Total rubber production is inferred from carbon black production, published by the Bureau of Mines, assuming rubber products are 33 per cent carbon black by weight, and 60 per cent of total rubber output goes to the automotive industry.

[40] From *Mineral Facts and Problems,* Bulletin 630, Bureau of Mines, 1965 edition.

[41] See footnote 39.

[42] *Ibid.*

[43] From *Mineral Facts and Problems.*

[44] *Ibid.*

As noted above, about 30 per cent of these inputs go to increasing the inventory of vehicles.

The recycle picture is mixed. About 6.8×10^6 tons of steel were recovered in 1965 from scrap auto, bus, and truck bodies.[45] It is extremely difficult to say exactly what happened to the remainder. Presumably some was lost during manufacturing processes, and probably most of this was returned to production as "new" scrap of unidentifiable origin. Essentially all the lead in storage batteries is ultimately recovered, and the majority of tires apparently are recapped at least once. About 0.3×10^6 tons, or 15 per cent, of all rubber is recycled, mostly old tires; however, a substantial tonnage of rubber—possibly 0.5×10^6 tons per year—is powdered. Some enters the air as organic dust that may be a health hazard. Ultimately it is mostly washed into rivers and streams; the remainder is probably burned or added to the growing inventory of old tires accumulating haphazardly in garages, on vacant lots, etc.

At present prices not all discarded automobile bodies can be utilized as open-hearth or oxygen-process scrap, owing to the presence of awkward contaminants, such as copper (~30 pounds per car), lead (30 to 40 pounds per car), aluminum (30 to 90 pounds per car), zinc (30 to 45 pounds per car), nickel, and chromium as well as nonmetallic impurities. Iron and steel constitute only about 79 per cent of the curb weight of an average automobile, and, even with tires, fluids, and batteries removed, about 15 per cent of the residual weight consists of materials which the potential scrap user does not want.[46] And the trend is toward increasing the unwanted materials. Thus, the annual supply of discarded motor vehicles (running at close to 6.5 million per year) is increasing faster than demand for this resource. The surplus piles up in automobile "graveyards" or as abandoned hulks scattered over the

[45] Testimony by W. E. Gilbertson, Public Health Service, 1966 air pollution hearings. Mr. Gilbertson estimated that this accounted for 94 per cent of the vehicles scrapped in that year, but this assumed an average weight of steel per vehicle (2,131 pounds) based on automobiles alone. In actual fact the average weight per vehicle must be somewhat higher, since about 12 per cent of the vehicles scrapped were trucks or buses, some of which were equivalent to several automobiles in weight. Hence, it might be fairer to assume 90 per cent of the vehicles were accounted for.

[46] This suggests the possible desirability of a disposal tax graduated to take account of the ease with which the materials contained in the automobile lend themselves to recycle.

countryside, or—when the concentration of eyesores reaches the threshold of toleration—the junk cars are sometimes incinerated and/or compressed and used in landfill operations. However, the main outlet in recent years has been to increase the inventory of junk cars in automobile graveyards.

Where a potential demand for scrap exists locally and the supply is large enough, new technology makes it possible to improve the quality of the scrap metal at an economic cost. The method utilizes giant shredders which automatically fragmentize the auto bodies (minus engines, radiators, and gas tanks) and magnetically sort the ferrous from the nonferrous materials. Combustible material is burned off, and the ferrous residue is heated and rolled to increase its density. The resultant product is equivalent in quality to desirable No. 1 heavy melting scrap (e.g., steel rails) in quality and degree of contamination. About 2.5×10^6 tons were processed in this way in 1965; unfortunately the capital investment required for such equipment is such that it can be justified only in densely populated areas—where, of course, the recovery operation itself is likely to impose external costs.

The processing required to return the solid wastes (junk autos) to the production cycle results in discharge of residuals to the air as contrasted with compression and sanitary landfill *without* incineration or dumping in shallow coastal waters, where the hulks can become convenient breeding grounds for marine fauna. At the present annual rate of motor vehicle discard, about 2.5×10^6 tons of nonmetallic materials must be disposed of. Assuming this is 80 per cent combustible, complete combustion would yield about 0.5×10^6 tons of ash—much of it, however, emitted to the atmosphere as particulates.

Whole Sector—Interdependencies

Chart 12 shows a schematic residual materials flow for the household sector, excluding automobiles. From it one can obtain a conception of the interdependencies of the various residuals streams. For instance, if liquid residuals are given high level treatment and if all sewage sludges and the maximum amount of garbage and refuse are incinerated, the majority of the final residuals load from households, which must be disposed of into the environment, is gaseous. On the other hand, if no incineration is conducted at all and sewage treatment is pressed to the

Chart 12. Household Residual Materials Flow (per capita)

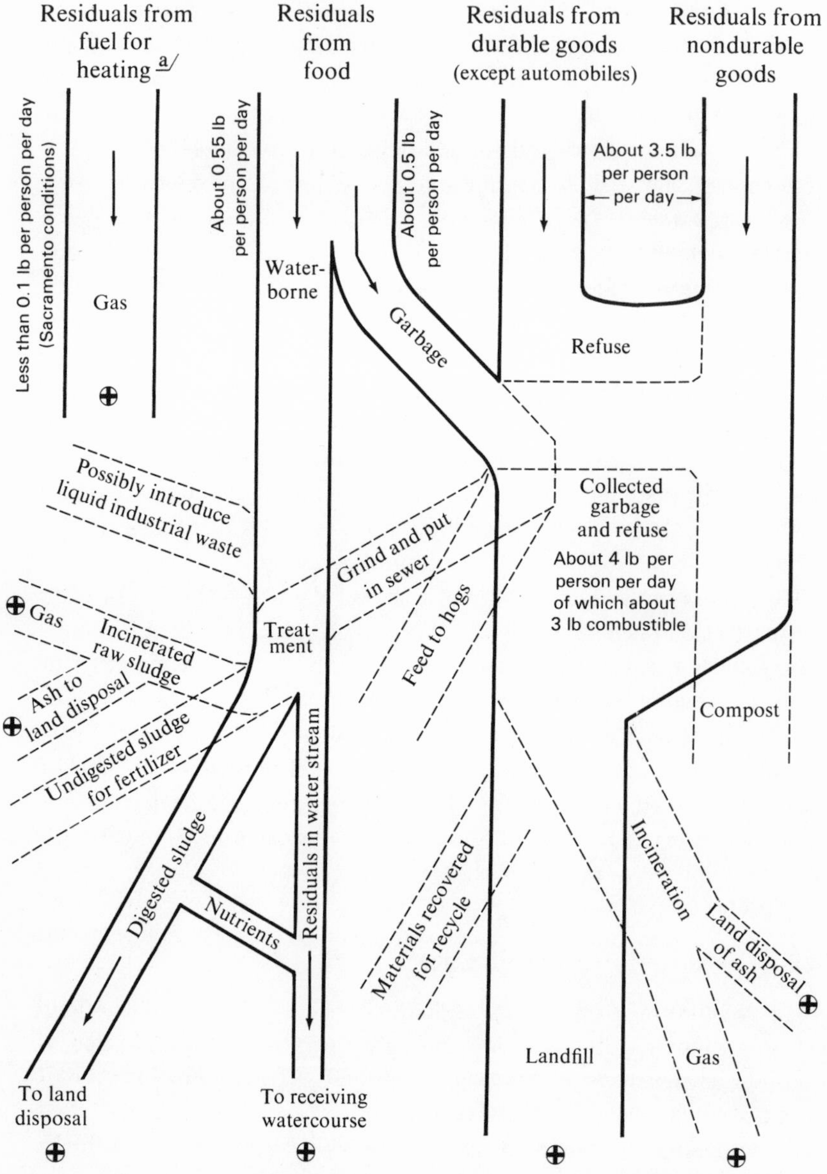

Residuals from fuel for heating [a]
Residuals from food
Residuals from durable goods (except automobiles)
Residuals from nondurable goods

Less than 0.1 lb per person per day (Sacramento conditions)
Gas
About 0.55 lb per person per day
About 0.5 lb per person per day
About 3.5 lb per person per day
Water-borne
Garbage
Refuse
Collected garbage and refuse
About 4 lb per person per day of which about 3 lb combustible
Possibly introduce liquid industrial waste
Grind and put in sewer
Feed to hogs
Treat-ment
Gas
Incinerated raw sludge
Ash to land disposal
Undigested sludge for fertilizer
Compost
Digested sludge
Nutrients
Residuals in water stream
Materials recovered for recycle
Incineration
Land disposal of ash
Landfill
Gas
To land disposal
To receiving watercourse

[a] *This stream is cut out if electricity is used. The waste stream then appears in the electric power sector.*

⊕ Potential for external cost
---- Alternative flows

point where all solids are removed from the liquid residuals stream, all but a few per cent (gases from space heating) of the weight of residuals flowing from households becomes a solid or semisolid. Finally, should garbage and refuse grinding and discharge to the sanitary sewers be practiced on a large scale and no liquid residuals treatment to remove solids be conducted, very little gaseous residual would flow from households and perhaps half the weight of residuals would be deposited into the aquatic environment. The remainder would be solids.

A general alternative to trying to "throw away" these residuals is to reincorporate them into the productive process, thus reducing the overall throughput of the production-consumption system. At some cost, virtually all the household residuals could be recaptured for useful purposes. Thus, sewage solids and wet garbage could be composted and recovered for fertilizer.[47] Metals and some other durable materials (including glass) can be, and sometimes are, recovered for reuse. Plastics pose perhaps the most difficult recycling problem. Still, application of known recovery technologies could reduce the net outflow of residuals from the household sector to a very low level.

We are not proposing that this is the appropriate solution. The cost of recovering the material may be, and often is, considerably higher than that of obtaining new materials from nature, even when external costs associated with disposing of residuals into the environment are taken into account. But, as outlined in Chapter I, there are reasons to think that the present system for organizing production (the market combined with rather ad hoc controls) does not produce the optimum degree of reuse possible with today's technology, nor does it stimulate an appropriate rate of improvement in that technology. Under current conditions few, if any, individuals and private or public agencies systematically bear the external costs of disposing of residuals into the environment. We rely (weakly) on scattered and more or less ad hoc controls. Moreover, we are not even equipped institutionally to look at the whole residuals management problem. Water pollution control is in one federal department, control of air pollution and solid wastes in a different department. Except in a few cases, such as those of Oregon's

[47] But transport of the fertilizer back to points of use would itself generate some residuals. Indeed, virtually all recovery operations require the application of energy (at a minimum), a factor which should not be forgotten in calculating their *net* impact on residuals.

Department of Environmental Quality, Pennsylvania's Environmental Protection Agency, and New York City's Environmental Protection Administration, responsibility is similarly splintered at state and local governmental levels. Restrictions relating to water pollution usually say nothing about air pollution or disposal of solids. Often the water pollution authority is pushed to protect watercourses by producing gases or solids which it "throws away" into the environment. We will not here pursue the matter of how the matrix of controls and regulations might be altered to improve the situation, but we do want to underline that arrangements which fully reflected all external costs would tend to lead to higher rates of reuse, less net throughput, and a more rapid rate of technological improvement in recovery systems.

Concluding Comment

We have now surveyed the flow of materials, recycle and by-product recovery possibilities, and residuals in several broad sectors of the economic system. We have seen the applicability of the materials balance approach to an understanding of residuals quantities, control options, and interdependencies among various forms of residuals. Our task in the following two chapters is to wed the materials balance concept to some models of the economy which have been used to examine general interrelationships in the exchange system and the welfare properties of various policy changes. We may also find that wedding the materials balance concept to simpler operational forms of the general equilibrium model can bring forth a useful method for projecting changes in residuals that result from increased demand for final products or from a changing bill of goods.

Appendix to Chapter II

WASTE HEAT AND NOISE—ENERGY RESIDUALS

In the United States in 1965, gross energy consumption from all fuels, including hydroelectric power, was very close to 4×10^{16} Btu or 1×10^{16} kilocalories.[1] Of this total it can be assumed roughly that all energy content in fuel used to generate electricity as well as virtually all electrical energy (except for a small portion used in electrolytic decomposition of aluminum, magnesium, sodium, and phosphorus compounds) is eventually reduced to heat. Similarly, most of the energy value of fuel used in transportation (24 per cent of the total) and for space heating is reduced to heat, except for unburned hydrocarbons, which account for about 2–2.5 per cent of the total. In the industrial category, the energy content of some petrochemicals is largely preserved in the final product, viz., synthetic rubber, carbon black, artificial fibers, plastics, lubricants, asphalt, and road oils. This category accounts for 5 per cent of the total energy content of all fuels. Moreover, a substantial part of the heat content of some fuels (mainly coke) is used to reduce metallic ores—mainly iron ore—from the oxide to the metal, or in the manufacture of lime and portland cement from limestone. The primary metals sector also accounts for 5 per cent of total energy (3 per cent iron and steel, 2 per cent other metals), of which possibly half is wasted. Altogether perhaps 90 per cent of the energy consumed in the United States is returned promptly to the environment as "waste" (i.e., irrecoverable) heat. In quantitative terms,

[1] Most data for the portions of this chapter dealing with waste heat are from Vogely and Morrison, "Patterns of Energy Consumption."

this now amounts to 3.6×10^{16} Btu per year and is increasing at about 3.3 per cent per year.

The great bulk of the waste heat is returned to the atmosphere. A comparatively small fraction ($\sim$15 per cent), mostly from electric utilities and large industrial users, is dissipated into rivers and streams (but most of this is eventually returned to the atmosphere by convection or evaporation and condensation), where under some circumstances it can cause substantial external costs.[2] Since the biochemical demand for oxygen (BOD) increases in proportion to temperature, waste heat discharges tend to decrease the assimilative capacity of watercourses. Also, elevated temperatures often tend to increase the sensitivity of local fauna to certain toxic substances which may be present in the water.

By far the largest amount of waste heat discharge to watercourses comes from thermal power plants. At present efficiency levels, about 6,230 Btu of heat must be dissipated by cooling water, per kilowatt-hour of electricity generated. This is the result of a fuel consumption of 10,700 Btu, of which some 10 per cent goes up the stack or is lost in a number of individually minor ways, about 3,400 is converted into electrical energy, and the remainder is transferred to the cooling water. Improvements in energy conversion efficiency (which have been occurring persistently over the years) reduce the amount of waste heat which must be dissipated. The advent of atomic reactors as heat sources for thermal power plants will, however, tend to increase the heat load per kwh which must be disposed of. This is presently a worrisome public policy problem.

Cooling towers can be and often are used to take the heat load off the stream (or reduce water intake) but only by transferring heat directly to the ambient atmosphere instead. The temperature of the water coming from the condensers is reduced by partial evaporation, the balance of the water being recirculated. Generally speaking recirculation is rather low in cost—usually less than 1 cent per thousand

[2] These may perhaps be partly offset by some external benefits. For example, a small percentage of the water delivered through municipal systems is heated and an increase in the temperature of intake water can save a certain amount of energy; the palatability of the water is reduced, however. Also winter sport fishing is sometimes improved below large heat discharges because the fish tend to congregate there, if they manage to survive the overheated summers.

Table A–1. Energy Costs from New Coal-Fueled Power Plants
(Mills per kwh)

	Once through cooling	Recirculated cooling	% increase
65% capacity factor:			
Capital costs	2.295	2.421	5.5
Operation and maintenance	0.295	0.311	5.5
Fuel at 25¢ per mBtu	2.104	2.127	1.1
Total	4.694	4.859	3.5
80% capacity factor:			
Capital costs	1.865	1.968	5.5
Operation and maintenance	0.261	0.275	5.4
Fuel at 25¢ per mBtu	2.104	2.127	1.1
Total	4.230	4.370	3.3

Size of generation unit: 1,000 megawatts
Cooling water requirements—once-through: 400,000 gallons per min.
New water (make-up) for recirculated cooling: 10,000 gallons per min.
SOURCE: Calculations based on *Costs of Large Fossil Fuel Fired Power Plants,* Jackson & Moreland, Inc., Boston, Mass., Apr. 30, 1966.

gallons.[3] The amount of water involved in large plants is, however, sufficiently great that aggregate costs can mount quite high. For example, total costs will be increased by 3–6 per cent depending upon whether we are speaking of fossil fuel or nuclear plants. (See Table A–1 for an illustrated calculation.)

At first it would appear that the discharge of waste heat to the atmosphere is of little significance (except locally), but recent evidence indicates that this may not be so. Cities have measurable and even rather large effects on their own climates in terms of differences in temperature, humidity, precipitation and fog frequency, and wind speed between the city and its environs. Not all the effects listed in Table A–2 are necessarily negative (for example, longer periods free of frost may be favorable), but others are, such as persistent smog and dust cover. Part of the difference in temperature between the central city and its suburbs, which is related to the other phenomena, is due to the tremendous amounts of heat generated by various energy

[3] See P. H. Cootner and G. O. G. Löf, *Water Demand for Steam Electric Generation* (Resources for the Future, 1965).

Table A–2. Differential Climatic Effects Between a City and Its Environs

Climatic effect	City/environs (per cent)	Difference (per cent)
Solar radiation (insolation) on horizontal surfaces	85	−15
Ultraviolet radiation, summer	95	−5
Ultraviolet radiation, winter	70	−30
Annual mean relative humidity	94	−6
Annual mean wind speed	75	−25
Frequency of calms	115	+15
Frequency of cloud cover	110	+10
Frequency of fog, summer	130	+30
Frequency of fog, winter	200	+100
Total annual precipitation	110	+10
Days with less than 0.2 inch of precipitation	110	+10

SOURCE: W. P. Lowry, "The Climate of Cities," *Scientific American,* Vol. 217, No. 2 (Aug. 1967).

Note: Environs assumed to be 100 per cent.

consuming activities in the city.[4] For example, a blanket of warm air over the city combined with rapid cooling of surfaces such as rooftops and streets in the early morning hours can contribute to a temperature inversion and thereby increase the concentration of gaseous residuals in the atmosphere.

The exact role of heat discharges in influencing the climate of cities has not been defined, but it appears to merit study. Limiting heat discharges would be difficult and costly. There is no technological means on the horizon which would permit generating electric power in large quantities without waste heat production (although continued modest increases in efficiency can be anticipated). The most plausible approach to limiting the external costs is to concentrate on minimizing the effects. Thus heated but otherwise clean water could be pumped out to sea or to the deeper waters of large lakes. It is even possible that this would have some beneficial effects. For example, depositing the warm waters into the cool depths of lakes might help to break up thermal layer stratification of the lake and thus improve its waste assimilative capacity, or pumping into deep ocean areas might produce upwellings which would bring plant nutrients to the surface and help improve

[4] See W. P. Lowry, "The Climate of Cities," *Scientific American,* Vol. 217, No. 2 (Aug. 1967), p. 15.

fishing. This type of program requires regional planning to determine the optimum locations of large thermal power plants not only with respect to local markets, but also from the standpoint of making maximum use of the thermal assimilation capacity of the environment and producing beneficial effects on the environment.

One of the most widespread effects of waste energy, producing very large external diseconomies, is noise. It is doubtful, however, that much illumination of this problem can be achieved by viewing it as an energy residual problem. In contrast to the warming of streams and the atmosphere, only tiny amounts of energy are involved. One can see this easily by reflecting on the great amount of noise a transistor radio can extract from a couple of small batteries. Thus noise is perhaps better considered as a problem in its own right rather than as a matter of managing waste energy. There are, however, some interrelations between handling of material residuals and noise. Cooling towers on buildings, if improperly designed, can result in substantial and adverse noise generation. Incineration of solid residuals may also generate sizable noise residuals. More generally the handling of material residuals involves the application of energy which is rarely if ever totally noise-free.

Chapter III

RESIDUALS, GENERAL EQUILIBRIUM, AND WELFARE ECONOMICS

Introduction

The development of models for analysis of interdependencies of various parts of the economy has occupied some economic theorists for a long time. Discussion of such general interdependency models has its origins at least as early as Quesnay's Tableau Économique in the eighteenth century.[1] Sophisticated general equilibrium models were developed in the nineteenth century by Leon Walras, who applied concepts of Newtonian mechanics to the exchange processes of a theoretical perfectly competitive economy.[2] Names of major economic theorists of more contemporary times have been associated with further development of general equilibrium theory. These include Pareto, Wald, Hicks, Arrow, Samuelson, Debreu, and McKenzie, among others.

Much of the effort of the contemporary theorists has gone into providing "existence" and "uniqueness" proofs for solution of the systems of simultaneous equations characterizing the interdependent exchange system. This literature is among the most esoteric in economics and has made use of some deep mathematical theorems—especially the "fixed point" theorems. The general result following from these efforts has been the proof that under a set of more or less restrictive assump-

[1] This was an attempt to show the circulation of goods among the different classes of society and is one of the earliest efforts to apply scientific methods to economic phenomena. The Tableau was first published in 1758.

[2] *Éléments d'Économie Politique Pure*, 1877, Jaffé translation (London, 1954).

tions,[3] private exchange in a static competitive economy will lead to a single set of prices (resources, intermediate products, and final products) and an associated distribution of the goods that the economy can produce which corresponds to a "Pareto optimum," i.e., a situation in which all gains from trading are exhausted so that one person cannot become better off without making another worse off.

Among the major assumptions underlying this result, of interest in the present context, is that all production functions of producers and all utility functions of consumers are independent. This means that the output of one producer does not enter directly as an argument into the production functions of any other producer or the utility functions of any consumer. Decisions of others will influence the decisions of a particular producer or consumer via the intermediary of the market, i.e., through their impact on prices.

In "common property" situations this assumption cannot, in general, hold. The main topic of interest in this chapter is the question: what can we say about policy when this assumption fails to hold pervasively rather than just in isolated instances, as has been assumed in most discussions of policy criteria relating to externalities?[4] When we raise such a question we are inevitably in the world of "general" equilibrium analysis rather than the "partial" equilibrium analysis which has characterized most of the work done by economists on externalities. This question must be raised if we regard externalities associated with residuals from production and consumption as a normal and inherent part of these activities. And we have argued in Chapter I that they must be so regarded.

The version of a general equilibrium model that we have chosen to work with in connection with pervasive externalities associated with residuals is a particularly simple one known as the Walras-Cassel model. Its relative simplicity is derived to a considerable extent from the fact that it assumes fixed technical coefficients of production so that no optimization problem arises in connection with combining inputs and outputs. While this aspect lets us derive certain theorems, it does, as we shall see later, inhibit us from analyzing the full range of pos-

[3] A convenient summary of the assumptions different authors have made is found in J. Quirck and R. Saposnik, *Introduction to General Equilibrium Theory and Welfare Economics* (McGraw-Hill Book Co., 1968).

[4] Recall the discussion of this in Ch. I.

sible responses to restrictions on the use of the environment for residuals disposal. We deal only with material residuals in the following discussion.

Basic Model

The Walras-Cassel general equilibrium model,[5] extended to include intermediate consumption, involves the following quantities:

resources and services	products or commodities
$\overbrace{r_1, \quad \cdots\cdots\cdots\cdots \quad r_M}$	$\overbrace{X_1, \quad \cdots\cdots\cdots\cdots \quad X_N}$
resource prices	product or commodity prices
$\overbrace{v_1, \quad \cdots\cdots\cdots\cdots \quad v_M}$	$\overbrace{p_1, \quad \cdots\cdots\cdots\cdots \quad p_N}$
	final demands
	$\overbrace{Y_1, \quad \cdots\cdots\cdots\cdots \quad Y_N}$

The M resources are allocated among the N sectors as follows:

$$\begin{aligned} r_1 &= a_{11}X_1 + a_{12}X_2 + \cdots + a_{1N}X_N \\ r_2 &= a_{21}X_1 + a_{22}X_2 + \cdots + a_{2N}X_N \\ &\cdot \\ &\cdot \\ &\cdot \\ r_M &= a_{M1}X_1 + a_{M2}X_2 + \cdots + a_{MN}X_N \end{aligned}$$

or

$$r_j = \sum_{k=1}^{N} a_{jk}X_k \qquad j = 1, \cdots M \tag{1a}$$

In (1a) we have implicitly assumed that there is no possibility of factor or process substitution and no joint production. These conditions will be discussed later. In matrix notation we can write:

$$\begin{bmatrix} r_1 \\ \cdot \\ \cdot \\ \cdot \\ r_M \end{bmatrix} = \begin{bmatrix} a_{11}, & \cdots & a_{1N} \\ \cdot & & \\ \cdot & & \\ \cdot & & \\ a_{M1}, & \cdots & a_{MN} \end{bmatrix} \begin{bmatrix} X_1 \\ \cdot \\ \cdot \\ \cdot \\ X_N \end{bmatrix} \tag{1b}$$

where $[a]$ is an $M \times N$ matrix.

[5] The original references are Walras (see n. 2 above) and G. Cassel, *The Theory of Social Economy* (New York, 1932). Our own treatment is based largely on R. Dorfman, P. Samuelson, and R. M. Solow, *Linear Programming and Economic Analysis* (New York, 1958).

A similar set of equations describes the relations between commodity production and final demand:

$$X_j = \sum_{k=1}^{N} A_{jk} Y_K \qquad j = 1, \cdots N \tag{2a}$$

$$\begin{bmatrix} X_1 \\ \cdot \\ \cdot \\ \cdot \\ X_N \end{bmatrix} = \begin{bmatrix} A_{11}, & \cdots & A_{1N} \\ \cdot & & \\ \cdot & & \\ \cdot & & \\ A_{N1}, & \cdots & A_{NN} \end{bmatrix} \begin{bmatrix} Y_1 \\ \cdot \\ \cdot \\ \cdot \\ Y_N \end{bmatrix} \tag{2b}$$

and the matrix $[A]$ is given by

$$[A] = [I - C]^{-1} \tag{3}$$

where $[I]$ is the unit diagonal matrix and the elements C_{ij} of the matrix $[C]$ are essentially the well-known Leontief input coefficients. In principle, these are functions of the existing technology and, therefore, are fixed for any given situation.

By combining (1) and (2), we obtain a set of equations relating resource inputs directly to final demand, viz.,

$$\begin{aligned} r_j &= \sum_{k=1}^{N} a_{jk} \sum_{l=1}^{N} A_{kl} Y_l = \sum_{k,l=1}^{N} a_{jk} A_{kl} Y_l \\ &= \sum_{l=1}^{N} G_{jl} Y_l \qquad j = 1, \cdots \mathrm{M} \end{aligned} \tag{4a}$$

or, of course,

$$\begin{bmatrix} r_1 \\ \cdot \\ \cdot \\ \cdot \\ r_M \end{bmatrix} = \underbrace{\begin{bmatrix} a_{11}, & \cdots & \\ \cdot & & \\ \cdot & & \\ \cdot & & \\ & & \end{bmatrix}}_{M \times N} \underbrace{\begin{bmatrix} A_{11}, & \cdots & \\ \cdot & & \\ \cdot & & \\ \cdot & & \\ & & \end{bmatrix}}_{N \times N} \begin{bmatrix} Y_1 \\ \cdot \\ \cdot \\ \cdot \\ Y_N \end{bmatrix}$$

$$= \begin{bmatrix} G_{11}, & \cdots & G_{1N} \\ \cdot & & \\ \cdot & & \\ \cdot & & \\ G_{M1} & & G_{MN} \end{bmatrix} \begin{bmatrix} Y_1 \\ \cdot \\ \cdot \\ \cdot \\ Y_N \end{bmatrix} \tag{4b}$$

We can also impute the prices of N intermediate goods and commodities to the prices of the M resources, as follows:

$$p_k = \sum_{j=1}^{N} b_{jk} v_j \qquad k = 1, \cdots N \tag{5a}$$

or,

$$[p_1, \cdots p_N] = [v_1, \cdots v_M] \begin{bmatrix} G_{11}, & \cdots & G_{1N} \\ \cdot & & \\ \cdot & & \\ \cdot & & \\ G_{M1}, & \cdots & G_{MN} \end{bmatrix} \tag{5b}$$

In order to interpret the X's as physical production, it is necessary for the sake of consistency to arrange that outputs and inputs always balance, which implies that the C_{ij} must comprise *all* materials exchanges including residuals. To complete the system[6] so that there is no net gain or loss of physical substances, it is also convenient to introduce two additional sectors, viz., an "environmental" sector whose (physical) output is X_θ and a "final consumption" sector whose output is denoted X_f. The system is then easily balanced by explicitly including flows both to and from these sectors.

To implement this further modification of the Walras-Cassel model, it is convenient to subdivide and relabel the resource category into tangible raw materials $\{r^m\}$ and services $\{r^s\}$:

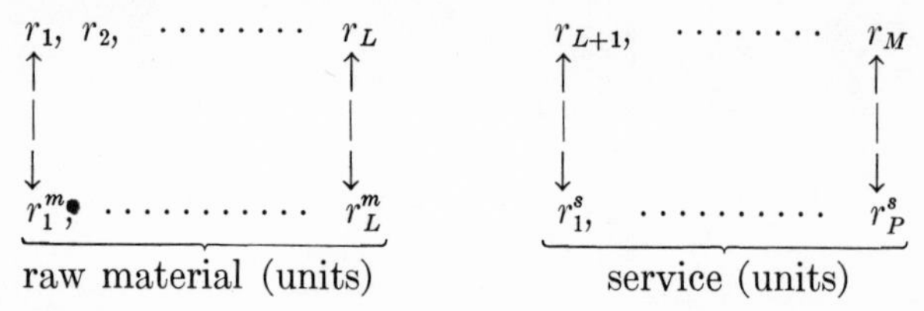

where, of course,

$$L + P = M \tag{6}$$

It is understood that services, while not counted in tons, can be measured in meaningful units, such as man-days, with well-defined prices. Thus, we similarly relabel the price variables as follows:

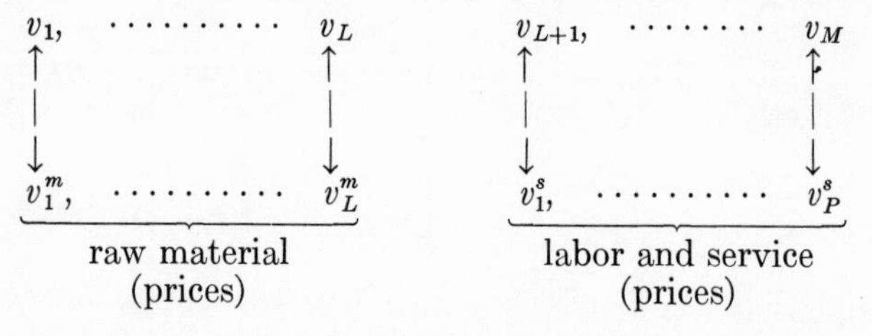

[6] We do not explicitly include the residuals-processing sector shown in Chart 1 of Ch. I.

The coefficients $\{a_{ij}\}$, $\{G_{ij}\}$ are similarly partitioned into two groups, e.g.,

$$\begin{array}{cccc cccc}
G_{1j}, & G_{2j}, & \cdots\cdots & G_L & G_{L+1,j}, & \cdots\cdots & G_{Mj} \\
\updownarrow & & & \updownarrow & \updownarrow & & \updownarrow \\
G^m_{1j}, & G^m_{2j}, & \cdots\cdots & G^m_{Lj} & G^s_{1j}, & \cdots\cdots\cdots & G^s_{Pj}
\end{array}$$

These notational changes have no effect whatever on the substance of the model, although the equations become somewhat more cumbersome. The partitioned matrix notation simplifies the restatement of the basic equations. Thus (1b) becomes:

$$M\left\{\begin{bmatrix} \cdot \\ \cdot \\ \cdot \\ r \\ \cdot \\ \cdot \\ \cdot \end{bmatrix}\right. \equiv \begin{bmatrix} r^m \\ \cdots \\ r^s \end{bmatrix}\begin{matrix} \}L \\ \\ \}P \end{matrix} = M\left\{\overbrace{\begin{bmatrix} \}L \quad G^m \\ \cdots\cdots\cdots\cdots \\ \}P \end{bmatrix}}^{N}\right. \begin{bmatrix} \\ Y \\ \\ \end{bmatrix}\Big\}N \tag{7}$$

while (5b) becomes:

$$\underset{N}{[p_1, \cdots p_N]} = \underbrace{[v^m \ \vdots \ v^s]}_{\substack{L \quad P \\ M}} \underbrace{\begin{bmatrix} G^m \\ \cdots\cdots\cdots\cdots \\ G^s \end{bmatrix}}_{N}\Big\}M$$

$$= [\cdot\cdot \ v^m \ \cdot\cdot] \begin{bmatrix} \cdot & G^m & \cdot \\ \cdot & & \cdot \\ \cdots & \cdots & \cdots \end{bmatrix} + [\cdot\cdot \ v^s \ \cdot\cdot] \begin{bmatrix} \cdot & G^s & \cdot \\ \cdot & & \cdot \\ \cdots & \cdots & \cdots \end{bmatrix} \tag{8}$$

The equivalent of (5a) is:

$$p^k = \underbrace{\sum_{j=1}^{L} G^m_{jk} v_m}_{\text{prices imputed to cost of raw materials}} + \underbrace{\sum_{j=1}^{P} G^s_{jk} v^s_j}_{\text{prices imputed to cost of services}} \quad \text{where } k = 1, \cdots N \tag{9}$$

We now wish to focus attention explicitly on the flow of materials through the economy. By definition of the Leontief input coefficients (now related to materials flow), we have:

$C_{kj}X_j$ (physical) quantity transferred from k to j

$C_{jk}X_k$ quantity transferred from j to k

Hence, material flows *from* the environment to all other sectors are given by:

$$\sum_{k=1}^{N} C_{\theta k}X_k = \sum_{j=1}^{L} r_j^m = \sum_{j=1}^{L}\sum_{k=1}^{N} a_{jk}^m X_k = \sum_{j=1}^{L}\sum_{k=1}^{N} G_{jk}^m Y_k \qquad (10)$$

using equation (1), as modified.[7] Obviously, comparing the first and third terms,

$$\underbrace{C_{\theta k}}_{\substack{\text{total material}\\ \text{flow } (\theta \text{ to } k)}} = \underbrace{\sum_{j=1}^{L} a_{jk}^m}_{\substack{\text{all raw materials}\\ (\theta \text{ to } k)}} \qquad (11)$$

Material flows to and from the final sector must also balance:

$$\underbrace{\sum_{k=1}^{N} C_{kf}X_f}_{\substack{\text{sum of all}\\ \text{final goods}}} = \underbrace{\sum_{k=1}^{N} C_{fk}X_k}_{\substack{\text{sum of all}\\ \text{materials}\\ \text{recycled}}} + \underbrace{C_{f\theta}X_\theta}_{\substack{\text{waste residuals}\\ \text{(plus accumulation)}}} \qquad (12)$$

Of course, by definition, X_f is the sum of the final demands:

$$X_f = \sum_{j=1}^{N} Y_j \qquad (13)$$

Substituting (13) into the left side of (12) and (2a) into the right side of (12), we obtain an expression for the residuals flow from the final product sector (f) in terms of final demands:

$$C_{f\theta}X_\theta = \sum_{j=1}^{N}\sum_{k=1}^{N} (C_{jf} - C_{fj}A_{jk})Y_k \qquad (14)$$

Material flows into and out of the intermediate product sector must also be in balance:

$$\sum_{j=1}^{L}\sum_{k=1}^{N} G_{jk}^m Y_k - \sum_{j=1}^{N} Y_j + \gamma \sum_{j=1}^{N}\sum_{k=1}^{N} C_{fj}A_{jk}Y_k = \sum_{k=1}^{N} C_{k\theta}X_\theta \qquad (15)$$

[7] Ignoring, for convenience, any materials flow from the environment *directly* to the final consumption sector.

The coefficient γ in the third term of identity (15) relates the proportion of recycled materials from the final sector, f which augments materials resources in the intermediate products sector. Of course, numerically, $0 \leqq \gamma \leqq 1$. If $\gamma = 1$, there is in effect no reduction in residuals flow due to recycling since recycling from the final sector appears as residuals in the intermediate products sector. If $\gamma = 0$, the total amount of recycling from the final sector never reenters the system as residuals flows, and consequently the implied mass balance between materials inflows and residuals outflows does not hold.

Total residuals flows to the environment are equal to residuals flows emanating from the intermediate products and final consumption sectors:

$$C_{t\theta}X_{\theta} = \sum_{k=1}^{N} C_{k\theta}X_{\theta} + C_{f\theta}X_{\theta} \tag{16}$$

By substituting (15) into (16) and (14) into (16), the following expression is obtained which relates total residuals flows to final demands:

$$C_{t\theta}X_{\theta} = \sum_{j=1}^{L}\sum_{k=1}^{N} G_{jk}^{m}Y_{k} - \sum_{j=1}^{N}\sum_{k=1}^{N} [(1-\gamma)C_{jk}A_{jk}]Y_{k} \tag{17}$$

Thus, total residuals flows from all sectors are related directly to the vector of final demands. If quite arbitrarily the types of raw materials are divided or aggregated in such a way that their number equals the number of final products (L is equated with N), the total residuals flow expression (17) simplifies to:

$$C_{t\theta}X_{\theta} = \sum_{j=1}^{N}\sum_{k=1}^{N} [G_{jk}^{m} - (1-\gamma)C_{jk}A_{jk}]Y_{k} \tag{18}$$

The implied mass balance relationship between basic material and residuals flows in the identities (11) through (17) is:

$$\sum_{k=1}^{N} C_{\theta k}X_{k} - (1-\gamma)\sum_{j=1}^{N}\sum_{k=1}^{N} C_{fk}A_{jk}Y_{k} = \sum_{k=1}^{N} C_{k\theta}X_{\theta} + C_{f\theta}X_{\theta} \tag{19}$$

That is, materials flows from the environment less continuously recycled products equals residuals flows from the intermediate products sector plus residuals flows resulting from final consumption.

Recycling is a dynamic or time related process since a portion of the final product is retained within the system for a specific time interval.

In deriving equation (17) we have made several implicit assumptions regarding how recycling enters the general equilibrium model: namely, that a proportion $(1 - \gamma)$ of tonnages recycled from the final to intermediate products sector is forever retained, and that the remainder of recycled final products (γ) appears as residuals from the intermediate products sector, neither of which contributes to additional tonnages of final product. While these assumptions definitely simplify the algebraic derivations, other more realistic assumptions could be made. For example, the $(1 - \gamma)$ component of recycled products could reenter the system as augmenting raw materials in equation (10) and the final product vector Y_k would increase. The increase would depend on the number of recycles permitted within one production period. Another approach would be to redefine each variable (and expand their number) in the model to include a time subscript, and then derive the optimal production and consumption plan over a finite planning horizon. In this case, recycled final products from the ith time period would augment raw materials available in the $i + 1$ period. In fact, recycling phenomena exhibit many similarities to neoclassical growth models where a proportion of current production is saved in order to augment the stock of capital and increase future levels of production.[8] However, recycling has the additional properties of reducing environmental degradation from residual wastes and environmental effects of raw materials extraction.[9]

The derivations could be simplified slightly if we assumed that there is no recycling per se. Thus, in the context of the model, we could suppose that all residuals return to the environmental sector,[10] where some of them (e.g., waste paper) become "raw materials." However, they would then be indistinguishable from new raw materials. In principle, this is an important distinction to retain.

The model just outlined demonstrates the "pervasiveness" of exter-

[8] See J. E. Meade, *A Neo-Classical Theory of Economic Growth* (London: Allen & Unwin, 1962), and F. H. Hahn and R. C. O. Matthews, "The Theory of Economic Growth," reprinted in *Surveys of Economic Theory*, Vol. II (New York: St. Martin's Press, 1967).

[9] Ralph d'Arge is working on these extensions under RFF sponsorship.

[10] In calculating actual quantities, we would (by convention) ignore the weight of oxygen taken free from the atmosphere in combustion and returned CO_2. However, such inputs will be treated explicitly later.

nalities associated with interrelationships between production, consumption, and environmental sectors when environmental (common property) resources are scarce and therefore have economic value but no price. Note, however, that we have treated flows in terms of conservation of mass and have not differentiated between relatively harmless and inert substances, such as soil and water, and biologically active and organic compounds or other substances with potentially more harmful aesthetic or health characteristics. However, theoretically, one could disaggregate the mass balance into types of residuals from each production process, so long as they summed to a total residuals flow which was consistent with the conservation of matter.

Also, we have not considered the problem of residuals flows changing environmental capacity and in so doing altering the availability of the physical environment. Also, one must consider the alternative possibility that materials resource extraction may reduce or increase the waste holding or assimilative capacity of the environment. For example, the consumptive use (conversion into products, or evaporation) of water reduces the waste assimilative capacity of watercourses. The effects of residuals accumulation over time may reduce (or at least alter) the availability of some natural resources. To take account of such possibilities, we add two more equations: one relating total residuals flows to the difference between environmental capacity $(C_\theta X_\theta)$ and raw materials flows; the second relating raw materials flows to the difference between environmental capacity and residuals flows. Of course, these two equations (or constraints) are not necessarily identical in that one would expect a priori that a ton of raw materials extraction would have a different impact on waste assimilative capacity than a ton of discharged wastes. In terms of our previous notation, these relationships can be expressed as follows:[11]

$$C_{t\theta}X_\theta \leq \beta^t(C_\theta X_\theta - \sum_{k=1}^{N} C_{\theta k}X_k)$$

and (20)

$$\sum_{k=1}^{N} C_{\theta k}X_k \leq \beta^j(C_\theta X_\theta - C_{t\theta}X_\theta)$$

[11] An implicit assumption of these relationships is that a solution exists if the equality signs are specified.

Inclusion of Externalities

The physical flow of materials between various intermediate (production) sectors and the final (consumption) sector tends to be accompanied by, and correlated with, a (reverse) flow of dollars.[12] However, the physical flow of materials from and back to the environment is only partly reflected by actual dollar flows, namely, land rents and payments for non-common-property raw materials inputs. There are three classes of physical exchange which are not fully matched by counterpart economic transactions; these are—

(1) private use of production inputs of "common property" resources, notably air, streams, lakes, and the ocean, which in addition may have an impact on the waste assimilative capacity of these resources;

(2) private use of the assimilative capacity of the environment to "dispose of" or dilute residuals;

(3) inadvertent or unwanted material inputs to productive processes —diluents and pollutants.

Normally these goods (or "bads") are physically transferred at zero price. This is not because they are not scarce in relation to demand— they often are in developed economies—or because they confer no service or disservice on the user—since they demonstrably do so— but because there exist no social institutions which would permit the resources in question to be "owned," and exchanged in the market.

The allocation of resources corresponding to a Pareto optimum cannot be attained without subjecting the above-mentioned nonmarket exchange to the moderation of a market or a surrogate thereof. In principle, the influence of a market might be simulated, to a first approximation, by introducing a set of shadow (or virtual) prices.[13] They may well be zero—where supply truly exceeds demand at zero price—or negative (i.e., costs) in some instances; they will be positive in others. Alternatively, we may think in terms of a set of environmental standards. These possibilities are discussed further below.

The total value of services performed by the environment cannot

[12] To be precise, the dollar flow governs and is governed by a combined flow of materials and services (value added).

[13] A similar concept exists in mechanics, where the forces producing "reaction" (to balance action and reaction) are commonly described as "virtual forces."

now be calculated, but it is suggestive to consider the situation if the natural reservoir of air, water, minerals, etc. were very much smaller, as if the earth were a submarine or "spaceship" (i.e., a vehicle with no assimilative and/or regenerative capacity). In such a case, all material resources would have to be recycled,[14] and the cost of all goods would necessarily reflect the cost of reprocessing residuals (and wastes) for reuse. In this case, incidentally, the ambient level of unrecovered wastes continuously circulating through the environmental inventory of the system (i.e., the spaceship) would in general be nonzero because of the difficulty of 100 per cent efficient waste-removal of air and water. However, although the quantity of waste products in constant circulation may fluctuate within limits, it cannot be allowed to increase monotonically with time, which means that as much material must be recycled—on the average—as is discarded. The bill of goods produced in a spaceship economy would certainly be radically different from that we are familiar with. For this reason, no standard economic comparison between the two situations is meaningful. The measure of worth we are seeking is actually the difference between the total welfare "produced" by a spaceship economy, where 100 per cent of all residuals are promptly recycled, and the existing welfare output on earth, where resource inventories are substantial and complete recycling need not be contemplated for a very long time to come. This welfare difference is undoubtedly very large, although we possess no methodological tools for quantifying it.

If the scarce common property environmental contributions (we will refer to them as environmental services) were paid for, the overall effect on prices would presumably be to push them generally upward. However, the major *differential* effect of undervaluing environmental services is that goods produced by high residual producing processes, such as paper-making, are underpriced vis-à-vis goods which involve more economical uses of basic resources. Thus, to avoid pervasive misallocations either a set of "market" prices must be simulated for environmental services or some other form of decision rules must be introduced. In the following section we consider the possibility of such a set of prices and/or rules either of an optimal or of "second best" variety.

[14] Any consistent deviation from this 100 per cent rule implies an accumulation of waste products, on the average, which by definition is inconsistent with maintaining an equilibrium.

Environmental Standards and Decentralized Decision-Making

Some General Notes on "Second Best" Problems

In previous sections, a model was outlined which demonstrates the "pervasiveness" of externalities associated with interrelationships between production, consumption, and environmental sectors. It is important to note that even if a particular type of productive activity does not directly utilize inputs from the environmental sector, it may do so indirectly through its demand for intermediate products from sectors that do. Thus, while selling insurance may not produce waste residuals directly, the use by this activity of paper means that insurance adds indirectly to demand for environmental waste residual disposal services. In fact, it is difficult to imagine any economic activity which does not directly and/or indirectly contribute to demands on the unpriced environmental sector. And if this is true, a nearly universal divergence (of greater or lesser degree) between prices and social costs is implied. This can be seen most readily if we observe that equation (5a) or (5b) above relates intermediate product prices (and final demand prices) to imputed prices of raw materials and services. Given the absence of a viable market for environmental services, the right side of (5a) or (5b) will not contain prices for these services, thus leading to an inequality between prices and social costs; that is, provided environmental services are scarce in the economic sense. The question addressed in this section is whether decentralized decision-making coupled with environmental planning on the part of a governmental unit can, in the presence of pervasive externalities, reestablish or approach an optimum in social product.[15] This is not to say that we imagine such an optimum as a pragmatic possibility. At most the highly abstract analysis we are able to conduct could yield reference points against which to compare alternative planning approaches. Specifically, we wish to know how desirable it might be to rely on decentralized decisions regarding the

[15] It must be emphasized that the Pareto conditions derived from decentralized decisions without consideration of environmental degradation are *not* optimal in the sense that *some* adjustment in allocation of resource inputs, and in production levels of final products, could make everyone better off without simultaneously making anyone worse off.

environmental sector where these decisions are tempered (or constrained) by collectively imposed standards.[16]

Before embarking on this discussion, we will undertake a brief review of several issues in the context of what economists call "second best" problems.[17] "Second best" versus "first best" is defined and related to the concept of Pareto optimality quite well by Davis and Whinston: "A Pareto optimum problem is one in which, given the market clearing conditions and the *technology* of the economy, all normative behavioral rules can be determined so that solutions of the system achieve a . . . maximum, [while] a second best optimum problem is one in which, given the market clearing conditions and the technology of the economy, at least one (but possibly more) of the behavioral rules is non-trivially specified, *cannot be changed*, and neither can the behavior of the deviant(s) be altered by *any* policy, and the remainder

[16] By standards, in this context, we mean broadly any type of mechanism for controlling the utilization of environmental services, e.g., taxes, subsidies, or direct control on emissions or quality deterioration. Generally, a particular level of control can be implemented in a variety of ways, provided the behavior of participants, such as firms, toward each type of controlling mechanism is predictable.

[17] The "second best" situation is usually cast in the following mathematical terms. Given some function $H(Y_1, \ldots, Y_n)$ that is differentiable and where partial derivatives are defined as H_i, second order partial derivatives as H_{ij} or H_{ii}, and where H is subject to a constraint J, with J_i and J_{ij} as previously defined for H, the maximum of H given J using the Lagrange multiplier method is: max. $\psi = H - \lambda J$ which yields the first order Paretian conditions $H_i - \lambda J_i = 0$ $(i = 1, 2, \ldots, n)$. If the additional restriction is added, $H_1/H_n = \rho\ (J_1/J_n)$ where $\rho \neq 1$, then the first order conditions become $H_i - \lambda J_i - \delta\ [(H_n H_{ji} - H_j H_{ni}/H_n^2) - \rho(J_n J_{ji} - J_j J_{ni}/J_n^2)] = 0 \quad (i = 1, 2, \ldots, n)$ where δ is a second Lagrange multiplier. Note, if there are m such additional restrictions, the first order conditions become

$$H_i - \lambda J_i - \sum_{j=1}^{m} \delta_m\ [(H_n H_{ji} - H_j H_{ni}/H_n^2 - \rho_m(J_n J_{ji} - J_j J_{ni}/J_n^2)] = 0 \quad (i = 1, 2, \ldots, n)$$

If the H and J relationship are separable in that $H = H_1(Y_1) + H_2(Y_2) + \ldots + H_n(Y_n)$, then in the case of only one additional restriction the first order conditions become:
$H_i - \lambda J_i = 0 \qquad (i = 2, \ldots, n - 1)$, and
$H_i - \lambda J_i - \delta\ [(H_n H_{1i} - H_1 H_{ni}/H_n^2) - \rho(J_n J_{1i} - J_1 J_{ni}/J_n^2)] = 0 \quad (i = 1 \text{ and } n)$,
since $H_{ij} = 0$, and $J_{ij} = 0$, for $i \neq j$. However, the materials balance model presented earlier illustrates that separability cannot be presumed to exist between *any* productive sector and the environmental sector such that first order Paretian conditions could be optimal with no adjustment or governmental intervention.

of the normative behavioral rules are to be chosen so as to achieve a . . . maximum."[18] Whether our problem falls within the realm of the "second best" by Davis and Whinston's definition depends upon the extent to which we must accept as immutable the *noninclusion of imputed* costs of environmental services in decentralized decisions. Conceptually, it may be possible to impute costs of utilizing the environment in such a way that a return to full Paretian conditions is possible. If so, the question arises: If because of institutional, administrative, or information cost restrictions environmental services nevertheless bear a zero price, are there other behavioral rules that can be superimposed to reduce or counteract these social-private cost discrepancies? There appear to be three methodological options:

(1) to presume that complete correction for all deviations is plausible to such an extent that the "first best" Paretian conditions can be attained through appropriate dosages of environmental standards, taxes, subsidies, or other policy instruments;[19]

(2) (or that) pervasity of externalities is so encompassing, and/or detailed information on the general equilibrium system so costly, that deviations between private and social costs from this source must be viewed as totally immutable;

(3) (or that) the deviations between social and private costs are only partially correctable, so that a "second best" in the Davis-

[18] See O. Davis and A. Whinston, "Welfare Economics and the Theory of Second Best," *Review of Economic Studies*, Vol. 32 (Jan. 1965). A divergence appears to have developed in the meaning of "second best" in the literature. While Davis and Whinston presume the behavior of the deviant(s) from Paretian conditions is immutable, and given this, ask how other participants (e.g., firms, individuals) might alter their behavior to achieve a "second best," McManus and Bohm, among others, view the "second best" problem as trying to correct for the deviant's behavior by "doing something about the deviant," be it a monopolist or groupings of technological and institutional externalities, as is analyzed here. Finally, Morrison has broadened the definition of "second best" theory to include replacement of old constraints, as well as adding new constraints to existing ones. M. McManus, "Private and Social Costs in the Theory of Second Best," *Review of Economic Studies*, Vol. 34 (July 1967); C. C. Morrison, "Generalizations on the Methodology of Second Best," *Western Economics Journal*, Vol. 6, No. 2 (1968).

[19] Davis and Whinston have suggested the informational content for such global corrections is "difficult to obtain" and in any case "in any real economy there will always be imperfections somewhere." See O. Davis and A. Whinston, "Piecemeal Policy in the Theory of Second Best," *Review of Economic Studies*, Vol. 34 (July 1967).

Whinston sense must be imposed following (or in conjunction with) the *partial removal* of deviations between social and private costs for environmental services.

In a welfare sense (disregarding information costs) the three options can be ranked. Case 1, of complete correction, allows the system to achieve the "first best" optimum. Case 3, while possibly not achieving the "first best" optimum, is less constrained than Case 2, so that the "second best" optimum achieved through appropriate governmental application of environmental standards would be at least as good in a welfare sense as could be derived under Case 2.[20]

A Simplified Model of the Economy

In this section we develop a simplified model of an economy[21] in order to study public policies directed toward ameliorating the social inefficiencies created by nonpricing of environmental services. As we note again in the final chapter, the model rests on drastically simplifying assumptions, and the conclusions we are able to draw from it can be considered no more than suggestive. This model takes as its starting point the materials balance general equilibrium model discussed previously. The model developed here assumes competition with consequent fulfillment of Paretian optimality conditions (prices equal mar-

[20] In saying "less constrained" we refer to the constraints imposed by assuming social and private cost gaps cannot be altered and thus are more binding than if these constraints could be relaxed, though not completely. Since the value of a constrained maximum cannot be less if one or more of the constraints is relaxed (though it might be constant), our conjecture on rankings must follow. Thus, whether one views pervasive externalities as natural or only transient and alterable facets of the economic system influences the type of, and level of achievement of, alternative governmental actions. However, of more importance to the discussion here is the realization that Case 1 may offer less social welfare if no corrections are made than either options 2 or 3. No correction for excess use of waste assimilation or removal would be expected to result in lower levels of social welfare than if a small degree of correction is imposed. This is not to argue that any amount of correction will bring about an increase in welfare, only that *certain* levels of correction brought about by imposition of environmental standards will. For a heuristic argument of the point regarding "small degrees of correction," see E. J. Mishan, *The Costs of Economic Growth* (London: Staples Press, 1967).

[21] Our model is an adaptation of a simplified economy-wide model used by Davis and Whinston for analyzing "second best" problems. Davis and Whinston, "Welfare Economics and the Theory of Second Best."

ginal social costs) everywhere in the economy except for environmental services. It also contains the assumption that all products are produced under conditions of constant costs; i.e., total production costs divided by physical production is constant regardless of the level of production, and the assumption is carried over from earlier sections that there is no substitution between factors of production.[22]

Given the following definitions:

ω_i = the utility function of the ith person, assumed to have arguments only of consumption directly by the person, exhibit global concavity, and where $i = 1, 2, \cdots, Z$.

$\bar{Y}_i$ = the total quantity of goods consumed by the ith person; i.e., a specific bundle of commodities.

y_{ik} = the quantity of product (final demand) of product k consumed by the ith person, where $k = 1, 2, \cdots, N$.

Y_k = production of the kth product available for final demands, where $k = 1, 2, \cdots, M$.

r_j^s = the total quantity of resource service j available for producing all final products.

$C_{jk}X_k$ = amount of material or residuals flow from sector j to sector k.

C_kX_k = a measure of capacity in sector k, either in units of mass, density, etc.[23]

$$\max.\ W = [\omega_1(\bar{Y}_1), \omega_2(\bar{Y}_2), \cdots, \omega_Z(\bar{Y}_Z)] \tag{21.1}$$

$$\sum_{i=1}^{Z} y_{ik} \leq Y_k \qquad k = 1, 2, \cdots, N \tag{21.2}$$

$$r_j^s \leq \sum_{k=1}^{N} G_{jk}^s Y_k \qquad j = L + 1, L + 2, \cdots, M \tag{21.3}$$

[22] The assumption of nonsubstitutability between factors including environmental services is a major impediment to the interpretation of our conclusions as regards realistic environmental policies. There is substantial evidence that substitution possibilities are widespread. See, for example, Löf and Kneese, *The Economics of Water Utilization*, p. 79.

[23] Note that this model is developed in terms of an assumed natural environment with overall fixed dimensions and with a capacity which is influenced only by resource extraction and residuals accumulation as specified in the constraint equations (20).

$$C_\theta X_\theta \geq \sum_{j=1}^{L} \sum_{k=1}^{N} G_{jk}^m Y_k - \sum_{j=1}^{N} \sum_{k=1}^{N} [(1 - \gamma) C_{jk} A_{jk}] Y_k = C_{t\theta} X_\theta \tag{21.4}$$

$$C_\theta X_\theta \geq \sum_{j=1}^{L} \sum_{k=1}^{N} G_{jk}^m Y_k = \sum_{k=1}^{N} C_{\theta k} X_k \tag{21.5}$$

$$C_{t\theta} X_\theta \leq \beta^t (C_\theta X_\theta - \sum_{k=1}^{N} C_{\theta k} X_k) \tag{21.6}$$

$$\sum_{k=1}^{N} C_{\theta k} X_k \leq \beta^j (C_\theta X_\theta - C_{t\theta} X_\theta) \tag{21.7}$$

$$\bar{Y}_i,\ Y_k,\ y_{ik} \geq 0 \qquad \text{for all } i \text{ and } k \tag{21.8}$$

These relationships simply state that we wish to maximize welfare of the economy as measured by the utility levels of persons 1 through Z, subject to the conditions that all markets for final goods are cleared and that no more resources, including environmental services, are used than are available. Following Davis and Whinston,[24] we reinterpret the utility bundles of goods $\bar{Y}_i$ as a scalar function (positively weighted vector sum) by attaching constant weights to each person's utility function and rewrite (21.1) as:

$$\text{max. } W = \sum_{i=1}^{Z} \alpha_i \omega_i (\bar{Y}_i) \tag{22}$$

For each constraint or grouping of constraints (21.1) through (21.8), there are nonnegative Lagrangean multipliers that reflect the cost of these constraints. Let us denote these as η_k for (21.2), μ_j for (21.3), K_t for (21.4), K_j for (21.5), λ_t for (21.6), and λ_j for (21.7).[25] These

[24] "Welfare Economics and the Theory of Second Best."

[25] Let $W^* = \sum_{i=1}^{Z} \alpha_i \omega_i^* (\bar{Y}_i^*)$ denote the optimal value of (22) given constraints (21.1) through (21.8); then for small changes in one of the market clearing constraints (21.2) there will be small shifts in the optimal value of all variables. Let us further assume the new equilibrium also satisfies the optimal conditions (23) through (29)

multipliers can be interpreted as prices in that they measure the change in welfare (measured in cardinal units of α_i) per unit change in the constraint. In a competitive market situation, excluding considerations of adjustment processes (either dynamic or comparatively static), the multipliers would equal prices of final goods, or costs of direct and indirect resource inputs, including environmental services.

The first order consumer optimum conditions for the above model are:

$$\alpha_i \frac{\partial \omega_i}{\partial y_{ik}} - \eta_k \begin{cases} = 0 \\ < 0 \end{cases} \quad \text{provided } y_{ik} \begin{cases} > 0 \\ = 0 \end{cases} \quad \begin{matrix} k = 1, 2, \cdots, N \\ i = 1, 2, \cdots, Z \end{matrix} \qquad (23)$$

which in verbal terms indicates that the ith person's marginal utility with respect to the kth final product, weighted by the interpersonal utility weight α_i, equals the price of product k if it is consumed, and must be less if the product is not consumed.

The market-clearing equations require that all markets for final products be cleared in order for equilibrium to be maintained, or if supply of final product exceeds demand, price becomes zero. Note that demand for final products cannot exceed supply because of the direction of initial inequalities in (21.2).

$$\sum_{i=1}^{Z} y_{ik} - Y_k \begin{cases} = 0 \\ < 0 \end{cases} \quad \text{provided } \eta_k \begin{cases} > 0 \\ = 0 \end{cases} \quad k = 1, 2, \cdots, N \qquad (24)$$

---

given in the text. Also, let b_k denote relaxation of one of the market-clearing constraints (21.2).

Then: $$\partial W^*/\partial b_k = \sum_{i=1}^{Z} \alpha_i(\partial \omega_i^*/\partial \bar{Y}_i^*)\,(\partial \bar{Y}_i^*/\partial y_{ik})\,(\partial y_{ik}/\partial b_k).$$

From constraint (21.2) the following shift is obtained:

$$\sum_{i=1}^{Z} (\partial \bar{Y}_i^*/\partial y_{ik})\,(\partial y_{ik}/\partial b_k) = 1 \qquad \text{for } j = k \text{ and zero if } j \neq k.$$

Multiplying this equation by η_i^* and subtracting it from $\partial \omega^*/\partial b_k$ above, one obtains:

$$\partial W^*/\partial b_k = \eta_k^* + \sum_{i=1}^{Z} [\alpha_i(\partial \omega_i^*/\partial \bar{Y}_i^*)\,(\partial \bar{Y}_i^*/\partial y_{ik}) - \eta_k^*\, \partial \bar{Y}_i^*/\partial y_{ik}]\,(\partial y_{ik}/\partial b_k).$$

But the square bracketed term equals zero by constraint (23), such that $\partial W^*/\partial b_k = \eta_k^*$. Note that η_k^* is dimensionally similar to α_i. If α_i is dollars per "util," i.e., the reciprocal of the marginal utility of money income, then η_k is dollars per unit quantity of Y_k.

Likewise, the Pareto conditions for resource services.

$$\sum_{k=1}^{N} G_{jk}^{s} Y_k - r_j^s \begin{cases} = 0 \\ < 0 \end{cases} \quad \text{provided } \mu_j \begin{cases} > 0 \\ = 0 \end{cases} \tag{25}$$

$$j = L + 1, L + 2, \cdots, M$$

The constraints on material flows and environmental services yield the following first order Pareto conditions:

$$\sum_{j=1}^{L} \sum_{k=1}^{N} G_{jk}^{m} Y_k - \sum_{j=1}^{N} \sum_{k=1}^{N} [(1-\gamma) C_{jk} A_{jk}] Y_k - C_\theta X_\theta \begin{cases} = 0 \\ < 0 \end{cases}$$

$$\text{provided } K_t \begin{cases} > 0 \\ = 0 \end{cases} \tag{26}$$

$$\sum_{j=1}^{L} \sum_{k=1}^{N} G_{jk}^{m} Y_k - C_\theta X_\theta \begin{cases} = 0 \\ < 0 \end{cases} \quad \text{provided } K_j \begin{cases} > 0 \\ = 0 \end{cases} \tag{27}$$

$$C_{t\theta} X_\theta - \beta^t \left(C_\theta X_\theta - \sum_{k=1}^{N} C_{\theta k} X_k \right) \begin{cases} = 0 \\ < 0 \end{cases} \quad \text{provided } \lambda_t \begin{cases} > 0 \\ = 0 \end{cases} \tag{28}$$

$$\sum_{k=1}^{N} C_{\theta k} X_k - \beta^j (C_\theta X_\theta - C_{t\theta} X_\theta) \begin{cases} = 0 \\ < 0 \end{cases} \quad \text{provided } \lambda_j \begin{cases} > 0 \\ = 0 \end{cases} \tag{29}$$

It is important to note that if $K_t > 0$, then $K_j = 0$, since recycling cannot by definition be negative, i.e., materials flows must be equal to or exceed residuals flows. Also, it would be expected that either $K_t = 0$ or $\lambda_t = 0$, and $K_j = 0$ or $\lambda_j = 0$. Thus, if (21.6) and (21.7) become effective constraints, the constraints on mass balance represented by (21.4) and (21.5) generally would become redundant. However, theoretically this need not happen, and so for completeness, (21.4) and (21.5) are included.

The environmental services constraints represented by (21.4) through (21.7) could be partitioned into types of environmental services, i.e., airsheds, watercourses, algae sinks, etc., provided the components summed to $C_\theta X_\theta$. The assumption is made here that environmental services are of a sufficient degree of homogeneity for a meaningful single set of prices to be obtained. For purposes of simplification, we will also assume the basic production unit in the environmental region is a particular sector k that produces one product Y_k. We note in passing, however, that this simplifying assumption, along with the above assumption of homogeneity of environmental services, would collapse in attempting to derive practical optimal policy prescriptions. For the model to provide exacting optimal environmental standards, each set of constraints would require disaggregation for individual firms with differing requirements on environmental media, and to take into account locational advantages and disadvantages.

Finally, the Pareto conditions for optimum production levels of each sector k are:

$$\eta_k - \sum_{j=L+1}^{M} \mu_j G^s_{jk} - K_t \frac{\partial C_{t\theta} X_\theta}{\partial Y_k} - K_j \sum_{j=1}^{L} G^m_{jk} - \lambda_t \left(\frac{\partial C_{t\theta} X_\theta}{\partial Y_k} + \frac{\partial \beta^t}{\partial Y_k} \right)$$
$$- \lambda_j \left(\frac{\partial \sum_{k=1}^{N} C_{\theta k} X_k}{\partial Y_k} + \frac{\partial \beta^j}{\partial Y_k} \right) \begin{cases} = 0 \\ < 0 \end{cases} \quad \text{provided } Y_k \begin{cases} > 0 \\ = 0 \end{cases} \tag{30}$$
$$k = 1, 2, \cdots, N$$

This last Pareto condition indicates that the price of the kth final good should be just equal to the marginal costs of producing it, where cost includes direct resource input costs to the kth product plus indirect costs of resource inputs used to produce other products necessary for production of the kth product, plus any costs of reduction in environmental assimilative capacity.

Now if the competitive economy were operating without cognizance of constraints (21.4), (21.5), (21.6), or (21.7), the first order conditions (30) for production would change to:

$$\eta_k - \sum_{j=L+1}^{M} \mu_j G^s_{jk} - K_j \sum_{j=1}^{L} G^m_{jk} \begin{cases} = 0 \\ < 0 \end{cases} \quad \text{provided } Y_k \begin{cases} > 0 \\ = 0 \end{cases} \tag{31}$$
$$k = 1, 2, \cdots, N$$

and (30) would be excluded from consideration. This means that producers consider only the costs of non-common-property raw materials and intermediate products in their production decisions.

So long as (21.4) through (21.7) are not binding, however, i.e., environmental services are of such magnitude as to be virtually free goods, no important difference would arise between the Pareto optimum outputs of Y_k specified by (30) or (31)—that is, assuming for the moment that residuals flows do not enter into the utility functions of individuals 1 through Z. However, once the assimilative capacity or other aspects of environmental media became relatively scarce the specification of optimal levels of final demands would change, as a divergence would arise between production conditions (30) and (31).

If the governmental agency responsible for establishing environmental standards has complete information regarding relationships (7) through (20), in principle it could establish relative tax payments or other environmental control measures so as to adjust (31) to (30) in each sector. The optimal tax on product k in equilibrium would be equal to:

$$K_t \frac{\partial C_{t\theta} X_\theta}{\partial Y_k} + \lambda_t \left(\frac{\partial C_{t\theta} X_\theta}{\partial Y_k} + \frac{\partial \beta^t}{\partial Y_k} \right) + \lambda_j \left(\frac{\partial \sum_{k=1}^{N} C_{\theta k}}{\partial Y_k} + \frac{\partial \beta^j}{\partial Y_k} \right) \tag{32}$$

$$k = 1, 2, \cdots, N$$

provided *all* raw materials including those drawn from common property environments are priced. If not, then those unpriced components of the fourth term in (30) would need to be included in (32).

Thus, the Pareto optimum conditions adjusted for pervasive externalities could be fulfilled throughout the economy, for both consumers and producers. If all the partial derivatives in (32) are constant, the taxes become proportional to the magnitude of final demands and the governmental agency's estimation of taxes is much simplified.

Thus far, we have only analyzed the case where pervasive externalities, induced by nonconsideration of environmental services, entered into the sphere of production activities. We now turn to the case where residuals flows simultaneously affect levels of welfare of each individual. This can be accomplished by including $C_{t\theta} X_\theta$ in the utility function of each individual. Note that this implies that environmental pollution, as residuals flows, is a "public good" in the Samuelson-Musgrave sense.[26] The first order conditions regarding the producer sec-

[26] R. A. Musgrave, "The Voluntary Exchange Theory of Public Economy," *Quarterly Journal of Economics*, Vol. 53 (Feb. 1939); P. A. Samuelson, "The Pure Theory of Public Expenditure," *Review of Economics and Statistics*, Vol. 36, No. 4 (Nov. 1954); P. A. Samuelson, "Diagrammatic Exposition of a Theory of Public Expenditure," *Review of Economics and Statistics*, Vol. 37, No. 4 (Nov. 1955).

tor specified in (30) are changed with the addition of the term below,

$$\sum_{i=1}^{Z} \alpha_i \frac{\partial \omega_i}{\partial C_{t\theta} X_\theta} \frac{\partial C_{t\theta} X_\theta}{\partial Y_k} \qquad (33)$$

$$k = 1, 2, \cdots, N$$

which states that the sum of the marginal disutilities of all individuals induced by an additional unit of residuals flow to the environmental sector must be added to gross resource prices for environmental services and prices of all other resources to obtain final product price. Therefore, market prices must not only reflect the scarcity of environmental services but also disutilities emanating from environmental degradation due to residuals flows.

The inclusion of (33) into the model greatly complicates the derivation of environmental standards, for now the government agency must have complete knowledge not only of the production aspects and material balances of the economy, and the price for a numeraire product, but also of the utility functions for each individual. In any case, we have shown that even if externalities pervaded the entire economy, if a government agency had the necessary information and was unconstrained as regards action, and no additional constraints on behavioral rules arose a full Paretian optimum would be possible. The government agency could, in principle, formulate a coherent and consistent set of environmental standards.

In the previous section we implied that if such a set of standards could be found, once implemented they would be viewed as "internal" within the system and individual decision-making could still be preserved. Here we have demonstrated that such a set of standards conceptually *does exist* and could be implemented via administered pricing by a government agency and still retain individual decision-making as regards markets for final products and utilization of resources by industry *other* than environmental services. Difficulties arise, however, in empirically determining the production relationships, materials balances, environmental assimilative capacities, and utility functions needed to establish the optimal set of environmental standards. It is very likely that costs of obtaining the requisite information would far exceed the gains from *exactly* achieving the Pareto conditions (26), (27), (28), (29), (30), and (33), at least in the near future. Thus we endeavor to analyze the two other cases posed earlier in this section:

to presume total pervasity of environmental externalities so that the deviations between private and social production costs, and/or benefits, are immutable, and the alternative that externalities are only partially immutable; i.e., a partial correction for residual flows to common property environments is possible. We now turn to look at these less theoretically satisfying, but possibly more relevant and interesting cases.

Complete or Partial Immutability and "Second Best" Policy

The question raised here is: Given *complete* or *partial* immutability in discrepancies between private and social costs due to pervasive externalities from residuals discharge, can a "second best" solution be found? In referring to *complete* immutability we are thinking of the case where it is impossible for the government to exercise *direct* controls over the use of environmental services by any sector or private industry. For example, the government could not impose taxes, or emission controls on waste flows of firms. In referring to *partial* immutability, we are identifying those cases where the government can impose taxes or controls on firms utilizing environmental services, but where these taxes are less than the *optimal* level developed in the last section. We also wish to know if such a "second best" solution impedes the function of voluntary exchanges in markets through individual actions.

Morrison has demonstrated that, without complete knowledge of utility functions and production capabilities (and materials balances in terms of our analysis) plus *complete* information on the form of private-social cost discrepancies, the "second best" solution is generally indeterminate.[27] Thus, the same or a greater amount of information is required to establish coherent and consistent environmental standards in the "second best" case, as is required in the "optimal" case examined earlier. This is an important finding for analyzing externalities, since it suggests that a completely coherent set of environmental standards may never be formulated. The costs associated with measuring all requisite data inputs are undoubtedly extremely high, and may exceed the gains from increased efficiency, at least for some level of accuracy.

[27] Morrison, "Generalizations on the Methodology of Second Best."

Let us assume that the discrepancies between private and social costs in production resulting from residuals discharge are immutable, in the sense explained above. Then, given the assumption of competition and constant cost technologies, each industry will fulfill the first order condition (31) of equating prices to marginal costs of production other than environmental services. If these nonoptimal conditions are viewed as unchangeable, then they become additional constraints on the system (21.1) through (21.8). An alternative in the "first best" sense would be to impose direct supply control constraints by government agencies on all residual producing industries. In doing so, however, we are *adding* a constraint to the system in addition to (31). The additional constraint very likely would conflict with constraints (31) and thus make irrelevant the "second best" case studied here, and it would remove entirely the possibility of voluntary market exchanges following its imposition. However, supply controls could be implemented in such a way that welfare would reach the "first best" optimum.

For the case of immutability involving (31), we are interested only in solutions where in equilibrium $\eta_k = P_k$, $k = 1, 2, \ldots, N$, i.e., the shadow price derived equals the market price, and we assume there is an inverse demand relationship relating the market P_k to demand for all products:[28]

$$P_k = f_k\left(\sum_{i=1}^{Z} y_{i1}, \sum_{i=1}^{Z} y_{i2}, \cdots, \sum_{i=1}^{Z} y_{iN}\right) \qquad k = 1, 2, \cdots, N \quad (34)$$

Since the government agency cannot adjust the first order conditions for any industry, by assumption it cannot influence directly the activities of industries in our example.

One conceptual alternative would be for the public agency respon-

[28] Negishi has shown that even though such a "perceived demand" relationship "...is presumed to be equivalent to the actual demand curve of an equilibrium, there exists no mechanism in the model which gives rise to this equivalence at the point of the second best solution. Unless the second best solution is known to the ... [industries] beforehand, it is impossible to perceive a demand curve like ... [our (34)] which is equivalent to the actual demand curve of the point of the second best solution." Negishi goes on to show that one can reformulate the perceived demand relationship in such a way that a consistent equilibrium for a single firm is obtained, but output in at least one sector of the economy is left indeterminate. T. Negishi, "The Perceived Demand Curve in the Theory of Second Best," *Review of Economic Studies,* Vol. 34 (July 1967).

sible for pricing of environmental services to develop public companies competitive with the private industry responsible for the bulk of utilization of environmental assimilative capacity. Through subsidies these public companies could eliminate those firms that did not take cognizance of their impact on the environment. We will not pursue this alternative, since it has been studied elsewhere,[29] and in our case with such a large number of industries contributing to total residual flows it appears extremely unpractical. The government agency's other option would be to adjust consumer demand to compensate for the nonoptimal behavior patterns of industry.

Taking into account equation (34) and the additional constraints given by conditions (31), an alternative set of first order conditions for consumers is developed, which replaces (23).

$$\alpha_i \frac{\partial \omega_i}{\partial y_{ik}} - \eta_k - \rho_k f'_{ik} \begin{cases} = 0 \\ < 0 \end{cases} \quad \text{provided } y_{ik} \begin{cases} > 0 & k = 1, 2, \cdots, N \\ = 0 & i = 1, 2, \cdots, Z \end{cases} \tag{35}$$

The first order conditions of production in equilibrium are identical to (3) except that a term $\rho_k f'_k$ is subtracted from the left side of the equality sign, where, in both (35) and the modified (30), ρ_k is a Lagrangean multiplier associated with the first order condition (31).

If constraint (31) is fulfilled, by definition:

$$H^k_\theta \equiv K_t \frac{\partial C_{t\theta} X_\theta}{\partial Y_k} + \lambda_t \left(\frac{\partial C_{t\theta} X_\theta}{\partial Y_k} + \frac{\partial \beta^t}{\partial Y_k} \right) + \lambda_j \left(\frac{\partial \sum_{k=1}^{N} C_{\theta k} X_k}{\partial Y_k} + \frac{\partial \beta^j}{\partial Y_k} \right) \tag{36}$$
$$= \rho_k f'_k \qquad k = 1, 2, \cdots, N$$

The adjustment to each consumer's demand for product k is functionally related to the demand for environmental services by producers of k, since

$$f'_k = \sum_{i=1}^{Z} f'_{ik} \tag{37}$$

[29] See Davis and Whinston, "Welfare Economics and the Theory of Second Best."

Therefore, we can rewrite (35) in equilibrium as:

$$\alpha_i \frac{\partial \omega_i}{\partial y_{ik}} - \eta_k - \frac{f'_{ik}}{f'_k}(H^k_\theta) \begin{cases} = 0 \\ \\ < 0 \end{cases}$$

$$\text{provided } y_{ik} \begin{cases} > 0 & k = 1, 2, \cdots, N \\ \\ = 0 & i = 1, 2, \cdots, Z \end{cases} \tag{38}$$

Here each consumer is taxed (or receives a subsidy) according to his purchases of products requiring environmental services. The greater a consumer's demand for products requiring relatively more environmental services, the larger is his tax (or smaller the subsidy). Note that f_{ik}/f_k can be rewritten in terms of price flexibilities (ϵ_k) as

$$\left(\frac{\epsilon_{ik}}{\epsilon_k}\right) \cdot \left(y_{ik} \Big/ \sum_{i=1}^{Z} y_{ik}\right)$$

Thus, taxes in this case would be functionally related to demand relationships for individual consumers as well as for markets and to the proportion of total demand accounted for by each consumer. In the special case where there is only one consumer, the consumption tax on each product k would equal the production tax derived earlier.

To summarize, consumer demands are adjusted through taxes and/or subsidies on final products to partially (or completely) reflect relative differences in requirements of environmental assimilative capacity of final products. For the government agency to be able to establish optimal or suboptimal (in the "second best" sense) consumer taxes, the agency must have information on the utility functions and market demands of all individuals. This complicates the derivation, in terms of informational requirements, of "second best" optimum taxes and subsidies as much as including residuals flows in the utility functions of each individual in such a way that residuals flows become a "public" good.

The last case we consider is where the government agency can partially adjust the discrepancy between private marginal cost and social marginal cost due to the zero-costing by industries of environmental services. In this case, the agency can apply taxes on polluting industries, but not at the optimum levels specified by the first order production conditions (30).[30] Thus, a partial removal of differences between private and marginal social costs is possible, but not such a complete adjustment that the price of each product is equated with its marginal cost, including costs of environmental services in each industry. Implicitly we are assuming there is some external rule or impediment which does not allow the government agency to price environmental services received by firms in accordance with the firms' use of such services.

In consequence, the government agency will have at least two actions open to it in attempting to achieve a "second best" in welfare for this case. First, it is able to *partially* adjust the costs of individual firms to reflect the social costs of environmental services embodied in each firm's product. Second, the agency through taxes and/or subsidies can make an adjustment to prices paid by consumers in order that consumer demand will more adequately reflect social costs of environmental services not taken into account by industries. We have already argued that through adjustments in costs of production to partially reflect the use of environmental services and through adjustments in taxes on products paid by consumers, a greater improvement in welfare is *generally* possible, in contrast to the case where all such adjustment must be accomplished through actions directed only toward consumers.

In this case of partial immutability of constraints on taxing industries, the government agency would need to intercede in final product markets (as regards demand curves "perceived" by firms and taxes) and in the direct and indirect use of environmental services by consumers. Of course, decentralized decision-making could be allowed to prevail once prices for environmental services and consumer taxes or subsidies were established. The prices or taxes (controls) would then function as natural impediments to the private enterprise system much as antitrust, crop acreage control, business practice, or "luxury" tax laws are viewed currently. However, the "second best" types of solu-

[30] An exposition in symbolic form of this hybrid case will not be developed here, but one can conceptually view the problem in a technical sense as follows: Constraints (30) contain a parameter that can be changed, but this parameter is bounded by zero and some positive proportion less than one of (32). The government agency then would select these parameters in conjunction with fulfilling the "second best" solutions by appropriate adjustment in the revised conditions (30) and for consumers via adjustments in conditions (23).

tions appear to complicate rather than simplify the rules for constraining "piecemeal" decisions on the part of government agencies and private decision-making. In an application of the "second best" rules in the "partial immutability" case here discussed, an industry would be required to observe demand relations established by government edict, and consumers would be taxed according to their demands on environmental services. Producers would also be taxed partially for their uses of environmental services. While decentralized decision-making could function within these constraints, it is doubtful that anyone would claim that decentralized decisions had not, for the most part, been emasculated.

Throughout our discussion, the assumption of fixed technical coefficients with regard to all resource inputs including environmental services, recycled products, and intermediate products has allowed us innumerable simplifications with regard to establishing optimal or suboptimal taxes for environmental services. With fixed coefficients, possible substitutions of capital and/or labor for raw materials extraction or residuals flows are excluded from consideration. Yet these substitutions may play an important role in adjusting the economy toward some maximum level of welfare given the qualitative and mass balance constraints for environmental assimilative and/or regenerative capacity. In general, such substitutions will increase the potential level of welfare because the economy has added degrees of flexibility in providing any desired and attainable flow of consumption goods. When such substitutions are available, however, the government agency must in essence have complete knowledge of all relationships of the economy and its citizens' desires and *then* must solve the gigantic problem of deriving optimal taxes, subsidies, and other controls for the entire economy with cognizance of the behavior of all firms and consumers.

A Recapitulation of Results of Previous Sections

Given the assumptions underlying our model, we were able to derive explicit environmental taxes on firms and for consumers so that the economy in principle would operate efficiently and the Paretian optimal conditions would be met by every sector.[31] This conclusion leads us to believe that if a government agency could procure the necessary information on materials balances and economic interdependencies for the entire economy, a coherent and consistent set of prices for environmental services could be determined. From our findings, it is our view that once such a set of environmental prices were imposed on firms or industries, these charges could become a part of the "natural" impediments to business enterprise comparable to other resource scarcities. Voluntary exchanges in markets and individual decision-making for the most part could be preserved. Of course, these conclusions are strongly predicated on the underlying assumptions of the model we selected to study, including the assumption of no noncompetitive firms or industries and the assumption that environmental charges could be efficiently implemented.

If residuals flows directly influence the level of well-being of consumers, then the government agency, in addition to knowledge of technical interdependencies and materials balances, must have knowledge of preferences of consumers in order to establish optimal environmental charges. The extent of public reaction to aesthetic degradations, smog, pesticides, fish kills, etc., in the recent past is indicative that consumer preferences must be taken into account in the formulation of prices for the use of environmental services.

We find that a set of coherent and consistent charges or prices is identifiable in the case where industry behavior resulting from the assumption that environmental services are virtually free is *immutable*. This nonoptimal but unchangeable behavior can be partially (or, in certain cases, completely) compensated by means of government regulation of consumer purchases through taxes and/or subsidies.[32] The

[31] The set of optimal prices for environmental services of course is determined by the amount of environmental services embodied in each final product including all residuals flows. If all technical relationships are of the fixed coefficient type, the two qualitative constraints (20) are linear, and residuals flows do not affect individuals' utility, then these charges or prices for environmental services can be directly added to final product prices with few complications. Otherwise, environmental charges may need to be related to the waste discharges of individual firms or consumers.

[32] This result is also predicted on the assumptions of fixed coefficients in production, linear environmental assimilative capacity constraints (20), and residuals flows not influencing individuals' utility.

taxes charged in this case were found to be functionally related to the amount of environmental services embodied in each final product. Once such taxes are imposed, markets can be allowed to function without controls, with individual choice and decision-making preserved. However, in this case of total immutability of industry's noncosting of environmental services, the government agency would need information on market demand relationships and individual utilities in order to establish an optimal set of taxes in addition to the informational requirements regarding materials balances and environmental assimilative capacity.

The final case studied was that in which environmental services can *partially* be charged to industries, but not at levels necessary to achieve Paretian conditions throughout the economy. In this case, the government agency would need not only to regulate residuals flows by industries through partial pricing of environmental services, but also to regulate consumer purchases by means of taxes and/or subsidies.

It is important to note that, once such environmental charges were implemented in all three cases studied, markets could remain relatively free in the sense that buyers and sellers could still establish prices. Also, once taxes were imposed on environmental services utilized by industries, these industries could still operate as private decision bodies with no additional manipulation by a government agency. Individual choice would still prevail as to the purchases of goods by consumers.

Undoubtedly the greatest impediment to developing optimal or "second best" control measures is the almost limitless amount of information required, which may well be the *most* binding constraint in that the complexity of pricing or tax rules, the cost of information, and the propensity to induce greater costs from information errors may at some point of accuracy overwhelm consideration of efficiencies analyzed here. Thus, in practice, comparatively crude rules of thumb would need to be devised. We come back to this matter in the final chapter.

Price and Welfare Impacts of Environmental Standards

Our next consideration is the impact of imposition of prices for environmental services on prices of intermediate or final products and prices of resource inputs other than environmental services, i.e., capital and labor. Conceptually, the imposition of prices for environmental services will have two distinguishable effects on prices: a general rise (reduction) in product (resource service) prices and a change in relative prices of both products and resource services. The increase in particular final product prices will be related to the total amount of environmental services embodied in each product, including not only the direct utilization of environmental services in the product's manufacture, but also indirect utilization through the use of other products which require environmental services. Thus, the production of certain chemicals requires the input of other previously processed products, and the total input of environmental services to chemicals includes not only environmental services directly used by chemicals but, in addition, environmental services required by the previously processed products.

Given explicit knowledge of equation (17) relating total residuals flows to particular levels of final demands, by displacing final demand by one unit (or dollar) for one final product, holding all other final demands constant, and repeating this procedure for each final product, total environmental service requirements per unit of each type of final product are calculable. Those final or intermediate products with relatively high environmental service requirements, if prices for environmental services are imposed, should exhibit higher prices than those final products with low or nearly zero residuals flows per unit of final demand. This discussion, however, has implicitly assumed completely inelastic changes in technology (as regards reduction in residuals flow) in response to changes in costs of residuals flow.

Scarcities of environmental services that are at all restrictive will raise costs of environmental services in relation to costs of other resource inputs. The change in relative resource prices may induce substitution of these resource inputs for environmental services which would tend to alter costs of inputs in relation to each other. The particular technologies imposed for illustrative purposes in previous discussions do not allow for such substitutions and thus imply (given a particular set of resource service prices) that the total impact of environmental charges will be transmitted through product price.

In terms of existing economic models, not much can be said regarding changes in welfare when more than one deviation (e.g., externality plus monopoly) pervades the economy, and only the externality is corrected for as in our preceding discussion. Even in the case of a single deviancy, questions of income distribution have not been satisfactorily included.[33] Thus, in our previous analyses and in the somewhat heuristic welfare argument which follows with regard to pricing of environment services, we must assume away both problems of monopoly and noncompetitive market structure and questions of income distribution.

Foster and Sonnenschein have recently been able to prove that a "radial decrease in distortion will unambiguously improve welfare," provided there is only one consumer, no products are inferior, and all products are produced at constant costs per unit.[34] By "radial decrease in distortion," Foster and Sonnenschein mean a uniform reduction in the absolute or percentage differences between a consumer's marginal rates of substitution and producers' marginal rates of transformation. Since our model contains a constant cost assumption,[35] and if consumers 1 through Z are similar in their preference for products and environmental services, we can infer from the above result that a radial decrease in distortion between prices and marginal costs will generally increase welfare. The degree of distortion associated with nonpricing of environmental services is equal, in our case, to H_{θ}^{k}, thus:

$$\psi\,(H_{\theta}^{k}) \qquad k = 1, 2, \cdots, N \tag{39}$$

$0 < \psi < 1$, imposed as a per unit tax on products 1 through k, would generally lead to an improvement in welfare. Note that the tax on each product is proportional to each product's demands on environmental services.

The result above regarding improvements in welfare, as Foster and Sonnenschein prove, is dependent to a large extent on the assumption of constant costs. If costs are not constant, or if preferences (and incomes) of individuals are dissimilar, the above conclusion regarding the change in welfare from imposing environmental charges may not be correct. However, from our heuristic argument regarding the probable

[33] E. J. Mishan, "Second Thoughts on the Second Best," *Oxford Economic Papers,* New Series, No. 14 (Oct. 1962).

[34] E. Foster and H. Sonnenschein, "Price Distortion and Economic Welfare," *Econometrica,* March 1970.

[35] That is if the partial derivatives in (32) are constant or, what amounts to the same thing, the relationships (20) between environmental assimilative capacity, residuals flows, and raw materials flows are linear in addition to all other production coefficients.

change in welfare brought about by imposition of a particular set of environmental charges, we believe that, once a materials balance is developed for a region, a quasi-coherent (not optimal, but probably welfare-improving) and consistent set of environmental charges can be developed.

By now the reader will certainly have become painfully aware that we have had to use extremely simplified models to analyze economic welfare issues associated with the emission of residuals from production and consumption activities in the economy. The analysis undertaken here must be regarded as only the barest start toward adequate analytical devices for dealing with the problem.

In some ways our discussion raises more questions than are answered, particularly with regard to aggregation and to viable definitions of the production unit or firm and environmental assimilative capacity.[36] We feel that the approach we have begun to explore provides a different perspective on the economics of environmental pollution than has been typical and one of potentially great relevance. It emphasizes that externalities are an inherent part of the production and consumption process and that consistent and coherent approaches to environmental quality management must take this into account. We hope other economists and other scientists will be stimulated to develop this approach further.

[36] However, within the general economic equilibrium framework we have been able to derive overall connections between the economy and the natural environment in which the economy resides through materials balance. In addition, simplifications notwithstanding, explicit valuations or shadow prices are derived for residuals flows altering environmental assimilative capacity (for example, killing bacteria which aid the decomposition of organic wastes by excess dumping of inorganic wastes into a watercourse) and for raw materials extraction which alters the availability of environmental capacity for assimilation of "residuals" flow.

Chapter IV

CONCLUSIONS, POLICY, RESEARCH

An Approach to Management of Residuals in a Region

In the foregoing chapters we have focused on materials and residuals flows at the national level where at least illustrative data could be readily developed. We then endeavored to link the concept of materials balance to conventional economic general equilibrium models. This effort necessarily proceeded at a highly abstract level, but we did use these models to investigate—at least in a preliminary way—some questions concerning the functioning of an economy characterized by "pervasive" externalities. In this final chapter we undertake to spell out what we take to be the implication of the above analyses for regional or "problem shed" management of residuals from production and consumption.[1] It is at the regional level that detailed management efforts must be undertaken. We also point to what we regard as some important areas of research in support of such efforts.

In the light of the findings of the previous chapters we suggest that something like the approach outlined in the following paragraphs is more likely to lead to desirable results than the present partial and

[1] A few examples of the sort of region we have in mind are the East Coast megalopolis, the San Francisco Bay area, Los Angeles–Orange County, Greater London and the Thames Estuary area, the Ruhr, the Saar, Upper Silesia, and the São Paulo region. Highly concentrated conurbations of this type do not constitute a large part of the earth's land area, but they do contain a disproportionate part of its population and economic activities and generate a lion's share of externalities.

ad hoc efforts. One component would be a more or less complete accounting of materials flow for the area, at least for the most important residuals generating activities and the most pervasive residuals.[2] This would involve some major complications, since regional economies are very "open." In other words, at least the more significant material imports and exports (including goods at various stages through the manufacturing process) would have to be accounted for. Getting these data, as well as those needed to estimate materials balances for various industries, would involve considerable primary data collection. But once the process were completed, a picture of materials and residuals flows in the area could emerge. Economic base-input-output models could then be used to project levels of activity and industry mixes into the future, as well as transportation demands, population, and other parameters of interest in constructing a new materials flow. In the first instance, present technology and existing levels of materials recycle and by-product recovery could be assumed. This procedure would differ from conventional approaches in that it would relate levels of residuals generation logically to the industrial and population base and account for all residuals in an internally consistent manner. Conventionally, emissions of air borne, liquid borne, and solid residuals are extrapolated separately.

As a further step, this procedure would permit analysis of the overall impact on all residuals of control measures instituted for one or more of them. As we have said in the introduction and elsewhere, for a given level of economic activity, a given efficiency of energy conversion, and a given degree of recycle and by-product production, reduction of one type of residual must come at the expense of creating another. Furthermore, residuals should be classified and quantified in terms of their potential recoverability as a basis for economic analysis of the latter alternative. Also, and quite important, the overall implications of projected changes in residuals handling (transportation, transformation, and disposal) and recovery technologies could be tested.

Another major effort would have to go into mathematically simulat-

[2] Any particular area will make a relatively small contribution to the levels of such "planetary" residuals as CO_2 and DDT. Policies with respect to the management of such residuals must be considered and implemented at a national and international level. But informational inputs from regional studies would be important in this connection.

ing the relevant natural system, especially the meteorological, hydrological, and ecological subsystems, in such a way that the time and spatial patterns of residuals concentrations, in probability terms, could be estimated, as well as secondary effects of residuals discharges. The latter would include photochemical reactions in the atmosphere, the reduction of dissolved oxygen in watercourses resulting from the discharge of organic residuals to them, and some other effects such as the ecological effects of nutrient discharges.

Tradeoffs among discharges to different environmental media, the costs of controlling discharges, the costs of recycle over and above those induced by internal profit considerations, as well as pertinent information of various types on the probable damaging effects of alternative concentrations and durations of residuals in the different environmental media would have to be considered. On this basis a reasonably consistent and coherent (even though not optional) set of standards might be established.[3]

Once established, the environmental standards would become part of the overall framework within which voluntary market exchanges take place. We have previously argued that, if these environmental standards—although themselves perhaps not fully optimal—are considered as fixed constraints on the system, the conditions of Pareto optimality can be met within them and be deemed to have the same normative significance as when achieved under the "natural" constraints of conventional resource scarcities, private ownership, and technological production functions. This result is very significant because it suggests that the role of decentralized decision-making with its inherent efficiencies could largely be preserved.[4] If control measures are instituted

[3] Among the many other difficult questions we do not address in this monograph is by what authority these standards would be set. What we have in mind is a decision by some representative political body charged with protection and enhancement of the environment. Bodies of the type we visualize already exist on a much more limited scale. The Delaware River Basin Commission is an example. An augmented Thames Conservancy in England and Ruhrverband in Germany are further examples of the type of regional environmental quality agency which might suit the purpose.

[4] Another area of great importance which we pass over is the role of government in implementing large scale environmental quality control systems in the region. Research on water quality management has shown clearly that such measures as reservoir regulation of river flows, mechanical aeration of residuals-receiving water courses, and treatment of municipal and industrial residuals together in collective regional treatment plants can enter efficiently into regional

in the country's various regions, then, once the economic system has adjusted to these "approximate" standards, the resulting relative prices might provide a firmer basis for "partial" damage estimation and refinement of control measures.

Immediate Next Steps

Our analysis suggests to us a need to move forward rapidly toward a fuller understanding of the economics, politics, and technology of residuals management. The best way to do this, in our opinion, would be to mount a well financed multidisciplinary research project to analyze residual flows as related to the present, and projected future, economic structure of a particular region. As explained above, this analysis should be combined with a study of the concentration and duration of the residuals generated in the various environmental media. Phenomena of degradation and dilution in the environment are influenced by stochastic processes in nature. Atmospheric wind patterns, temperatures, and temperature inversions are examples, as are variations in streamflow. Moreover, there are complex interrelationships between the biological phenomena, important in the degradation of residuals, and these physical phenomena. For example, the basic equations of the "oxygen sag" in residuals-receiving water include temperature as a major variable. Accordingly, close integration of studies in the various natural sciences as well as their integration with the economic studies would be needed. Among other things such a study would help reveal more clearly what are the major gaps in our knowledge.

There is now under way at Resources for the Future an effort to devise an operational analytical model for regional studies of overall residuals management. This model weds environmental diffusion models to a nonlinear economic optimization model and will be reported in a subsequent monograph by Clifford Russell and Walter Spofford.

Institutional Obstacles

We are not in a favorable institutional posture with respect to the type of approach we have discussed in this paper—in regard neither to

water quality management systems. See Davis, *The Range of Choice in Water Management*. Similar points could be made with respect to other environmental media.

analysis nor to implementation. We suggest that reorganization and consolidation at the federal level should seriously be considered. This would involve the creation of a new environmental agency to take over functions now scattered in the Department of Health, Education, and Welfare, the Department of the Interior, the Department of Commerce, the Department of Housing and Urban Development, the Department of Transportation, and other agencies.[5] For implemention of programs, regional environmental management agencies of some type will, we believe, have to be developed in the relatively near term future.

The regional case study we proposed should be suggestive of the type of regional management agency needed. In fact, study of institutional and organizational questions should be built into the project. Obviously, present governmental arrangements are not conducive to the execution of the regional study. A special group would have to be authorized and funded by Congress or by a private foundation or both.

The Challenge to Research

Introduction

We conclude this monograph with a discussion of what we regard as the most urgent areas of needed research if the materials balance–general equilibrium approach is to be made a useful tool for planning, policy analysis, and the implementation of residuals management programs. We make this discussion brief and general, since a detailed discussion of needed research, especially for regional analysis, will be one of the central elements of the Russell-Spofford report mentioned earlier. Here we merely aim to give some perspective on broad areas of needed research. In the following discussion we generally follow Chart 13 counterclockwise from the upper left. The reader may find it useful to refer to it from time to time.

Materials Flow

Our knowledge of the flow of materials through the economic system and their loss or purposeful discharge to the environment is extremely

[5] After this book had been set in type the President proposed such a reorganization which would create an "Environmental Protection Agency."

Chart 13. Schematic Depiction of a Residuals Management System

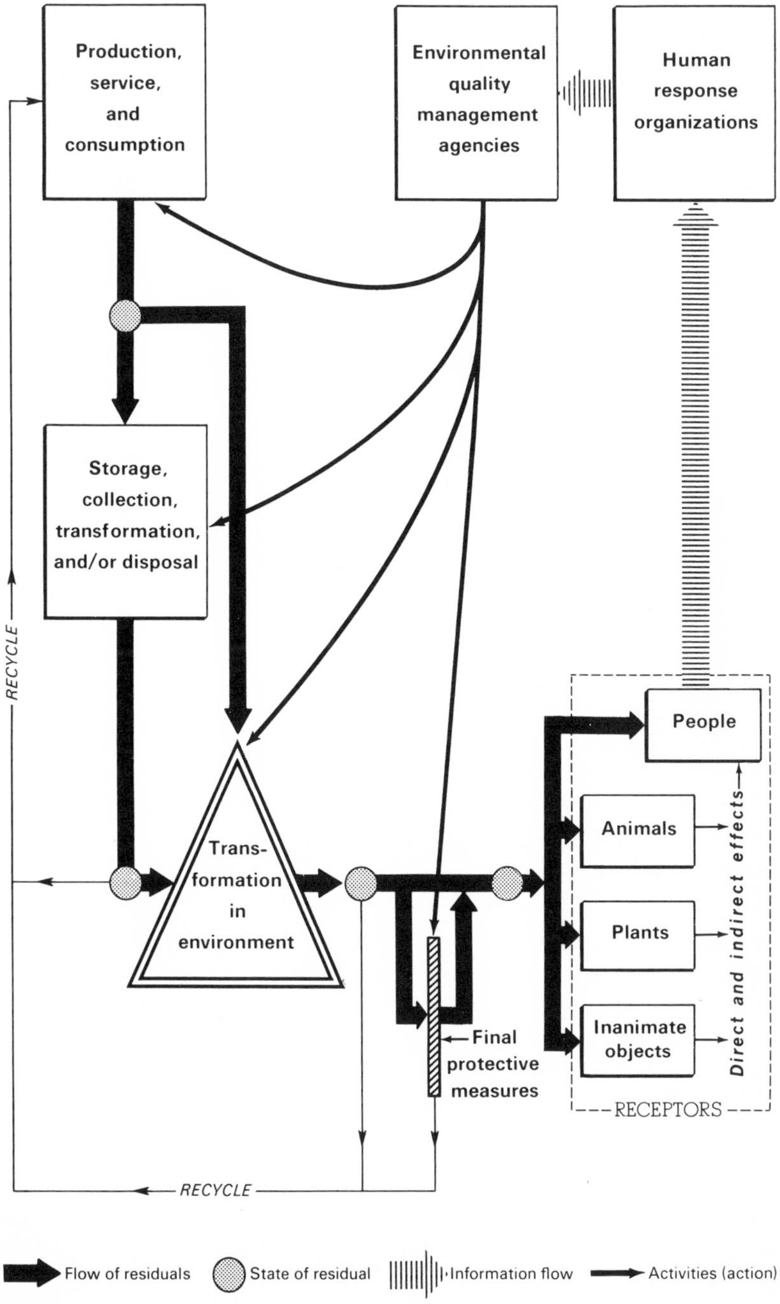

Production, service, and consumption

Environmental quality management agencies

Human response organizations

Storage, collection, transformation, and/or disposal

Trans-formation in environment

Final protective measures

People

Animals

Plants

Inanimate objects

Direct and indirect effects

RECEPTORS

RECYCLE

RECYCLE

Flow of residuals

State of residual

Information flow

Activities (action)

limited, especially with respect to industry. Studies of particular industries, similar to that of the beet sugar industry reported in Chapter II, could be very useful in implementing the materials balance approach. It is also possible that information gathered for the Census of Manufactures would be useful in establishing materials balances for particular industries. Learning about the opportunities for and costs of recycle and by-product recovery would, however, require careful study of individual industries. Some studies of this type are now succeeding.

While materials flow in the transportation and household sectors poses somewhat less diverse problems, studies of materials flow and residuals generation from these sectors should also be conducted. These should focus on alternative types of transportation systems and types of households. Of interest in the latter connection would be the difference between single family and apartment dwellings and the relationship of materials flow and residuals generation to family income, cultural background, place of residence, climate, etc.

A special study of the recycle (scrap) industries themselves could well be instructive. It is possible that through government incentives or assistance their effectiveness could be improved.

Interindustry Models

Models of interindustry economic exchange and materials flows are needed. In concept, these would be somewhat like the model developed in Chapter III. However, they should be a set of interlinked regional models for a fine enough grid that regions could be assembled from them for the purpose of environmental quality analysis. We realize that this may be an impossible aspiration, but we feel that at least feasibility studies for such a system of interlinked input/output materials flow models should be undertaken.

Residuals Handling Processes

There has been a tendency to view residuals handling (transportation, transformation, and disposal) processes in a narrow context. In liquid residuals treatment, performance and cost are usually measured in terms of the removal of degradable organics and suspended materials from the waste water stream. The fate of the removed material is seldom given much attention. Similarly, in treatment processes for air streams, exclusive attention is normally directed toward the removal of a few materials such as SO_2, NO_2, and particulates. Studies are needed which view residuals handling processes as systems for substituting one form of substance contained in the residuals stream for another form of residual or reusable product. Explicit attention must be given to such tradeoffs, rather than a tunnel vision focus on "reducing" individually gaseous, liquid, or solid residuals.

Projecting Residuals

We need to devise means of projecting the residuals generated and the residuals load which different rates and patterns of economic development, population change, and the location of both will impose on the environment. These should not be the conventional extrapolations of historical trends but should take account of interindustry relationships and materials manufacturing process interrelationships. A model incorporating various processing steps explicitly would permit forecasts to reflect the impact of technological changes in production as well as changes in final demand. The projection of residuals should incorporate the most advanced developments in technological forecasting. In particular, substantial consideration needs to be included of the trends in —and implications of—product outputs, nature of raw materials, and production processes.

Models of the Natural Environment

The natural environments of air, water, land, and their associated biological communities and chemical processes are characterized by complexity and more or less random change. There are models of the diffusion and degradation of substances in these environments but for the most part they are highly simplified and limited in the range of phenomena they can incorporate. Also, they are difficult to use except under simplified "steady state" conditions. We believe that a large-scale effort is required to improve our understanding of the natural systems and that mounting such an effort is limited not only by current shoestring financing but by arbitrary divisions among agency missions and, more fundamentally, among the relevant natural sciences. This work might well be a suitable focus for a large research organization.

Linkages to Affected Receptors

Incredible complications occur in linking concentrations and durations of substances in the environment to effects on receptors. Even something so relatively simple as corrosion has defied being linked functionally to the presence of substances in the environment thought to cause it. Vastly more complex are the processes of biological magnification and other ecological effects through food chains. Effects on human morbidity, mortality, genetics, and psychological reactions are only dimly perceived, although some encouraging work is under way. Factors related to aesthetics are only a little more understood. Again, this is a vast and difficult but exceedingly important area of research.

Social Damages

In the final analysis our concern with environmental quality stems from its short and long term effects on people and the things they value. Consequently, an important aim for research is to help discover what attributes of the environment people value and the factors which are likely to govern changes in these values in the future. These are the natural areas of concern of economics and social psychology. "Willingness to pay" for changes in environmental attributes is the standard measure of the economist. Its strength lies in that it provides a common measuring rod (dollars) and that, given that everyone operates under a budget constraint, it forces an automatic consideration of alternative ways of using resources. Its usefulness is limited in the presence of pervasive market imperfections (such as externalities) in that the market cost of alternatives for which the consumer's budget could be spent does not reflect their social costs. Thus, when the economy is subject to gross market failures, little normative significance can be accorded willingness to pay for a particular environmental change unless one regards the status quo of resources allocation as otherwise fixed. Also, when environmental (or other) changes are large enough to significantly affect a person's real income, the willingness-to-pay measure loses its status as a measure of social worth. Additionally, we may often be dealing with effects from the environment which are so subtle that the receptor does not understand the effects on him. Chronic health effects and contributions to generally stressful situations may fall into this category. Finally, perhaps the most important way that the individuals (families, firms) can adjust to adverse environmental circumstances is to change locations. Where this possibility is restricted (by racial prejudice, for example), it cannot be claimed that an optimal adjustment is possible even given the individual's income constraint. Thus, his willingness to pay may be distorted and the distribution of environmental quality may be an important policy consideration in its own right. Evaluating effects of environmental changes on human beings in a manner useful for policy planning, and management programs, is one of the most urgent research tasks we confront. This area demands theoretical and empirical research.

Institutional Arrangements

No amount of natural or social science research will help us deal with our environmental problems unless we learn how to implement effective management programs through legal and political institutions. How can we provide for a framework of incentives to compensate for powerful institutional interests in the status quo? How can we arrange institutions which comport with the diverse regional boundaries of the technological and economic aspects of environmental problems, whose patterns of representation press toward achieving the desired ends of man to the maximum extent, and which at the same time meet legal and political criteria of justice and equity? Surely these are the most difficult and least understood issues we have yet raised.

The Challenge to Economic Theory

Perhaps it is fitting that in a book which is primarily about the application of economics to an urgent national problem we should conclude our discussion of research needs with a challenge to economic theory.

We see an urgent need to develop more relevant and operational economic models for dealing with pervasive and subtle externality phenomena. A few economists have observed that external diseconomies increase rapidly (nonlinearly) and become more pervasive with economic and population growth, but comparatively little has been done to formulate analytical or normative models based on this insight. Assuming independence of production and utility functions in our basic economic models tends to become less defensible. The simplifying assumptions necessary in the model development of Chapter III, that is, of separability in environmental services and utility functions, of nonjoint production, and of simplified aggregation over interfirm, complementary-competitive relationships, are examples of the lack of realism in most current economic analyses. Also, the models we have developed are static and cannot take account of the accumulation over time of some of the more significant residuals discharged into the environment. Our theories of optimum rates of use of natural resources over time take no account of residuals discharge to unpriced sectors and nonoptimal rates of recycle of used or by-product materials. In our opinion, economic theorists face no more urgent task than to devise improved models for the analysis of environmental pollution, urban congestion, landscape deterioration, and the host of other externality (adverse environmental quality) phenomena which accompany economic growth.

Postlude—A New Malthusianism?

We have taken a look at an urgent national problem largely from the perspective of a particular discipline. It is the one that people at one time saw fit to call "the dismal science." Malthus, Ricardo, and many followers until contemporary times saw man engaged in a (losing) race against poverty, famine, and other grim population checks brought on by the pressure of numbers against the limited resources of soil, forest, and mine. Recently, we (at least in the United States) have gained confidence. Technology races forward, and of all the rates of increase futurists are pleased to project, this seems to be the fastest. Lower grades of ore, higher yields per acre, atomic blasts to enhance gas production, minerals from the deep sea—all of these are possible components of our salvation.

But are we raising the spectre of a new Malthusianism in this book?

Is our race now against increasingly stressful and impoverished environments, environmental controls which strive to deal with one problem but innocently cause another, increasing incidence of chronic disease, genetic damage (there are ominous hints from here and there), and finally, oxygen exhaustion? We don't know. But contemplate for a moment a world with twice or three times its present population (which it may have by the end of the century) and with the poorer countries producing and consuming at a rate more like that found in the industrialized countries today. We shudder and think there can be no more important tasks for contemporary science than to better understand the natural environment and man's relations with it.

But we must, as befits our dismal science, end on a dismal note. One would think that governments, especially in the wealthy nations, would be mounting substantial programs of basic and applied research to help the nations understand the consequences of their affluence for the natural environment. Or failing this, that large injections of research funds would be forthcoming from the great foundations. What we observe instead is an increasing preoccupation with the very short run brought on by the social and political crises of the day. Are we worrying about tactics when strategy is the most urgent consideration? Are we trying to fine-tune a system that is getting more and more grossly out of focus?

SOURCES FOR CHARTS

Chart 2: An unpublished research report by Richard J. Frankel, formerly of Resources for the Future, Inc., August 1967.

Charts 3 and 12: R. U. Ayres and A. V. Kneese, "Pollution and Environmental Quality," in H. S. Perloff, ed., *The Quality of the Urban Environment* (Resources for the Future, 1969).

Chart 4: "Chemical Origins and Markets," Stanford Research Institute, 1967. Chart prepared by R. U. Ayres from SRI Chemical Economics report data (in a completely different form).

Charts 5, 6, and 7: G. O. G. Löf and A. V. Kneese, *The Economics of Water Utilization in the Beet Sugar Industry* (Resources for the Future, 1968).

Chart 10: U.S. Public Health Service estimates, and *Chemical Week*, May 30, 1964.

Chart 11: Aerojet-General Corp., *California Waste Management Study*, August 1965.

RFF BOOKS ON ENVIRONMENTAL QUALITY

WATER TRANSFERS: ECONOMIC EFFICIENCY AND ALTERNATIVE INSTITUTIONS. *L. M. Hartman and Don Seastone*. 1970. 144 pages. $5.75.

THE QUALITY OF THE URBAN ENVIRONMENT: ESSAYS ON "NEW RESOURCES" IN AN URBAN AGE. *Harvey S. Perloff, ed.* 1969. 344 pages. $6.50 paper.

WATER MANAGEMENT INNOVATIONS IN ENGLAND. *Lyle E. Craine*. 1969. 130 pages. $3.50 paper.

THE RANGE OF CHOICE IN WATER MANAGEMENT: A STUDY OF DISSOLVED OXYGEN IN THE POTOMAC ESTUARY. *Robert K. Davis*. 1968. 204 pages. $7.00.

MANAGING WATER QUALITY: ECONOMICS, TECHNOLOGY, INSTITUTIONS. *Allen V. Kneese and Blair T. Bower*. 1968. 338 pages. $8.95.

THE WATER RESOURCES OF CHILE: AN ECONOMIC METHOD FOR ANALYZING A KEY RESOURCE IN A NATION'S DEVELOPMENT. *Nathaniel Wollman*. 1968. 296 pages. $7.50.

THE ECONOMICS OF WATER UTILIZATION IN THE BEET SUGAR INDUSTRY. *George O. G. Löf and Allen V. Kneese*. 1968. 134 pages. $4.00 paper.

THE ORSANCO STORY: WATER QUALITY MANAGEMENT IN THE OHIO VALLEY UNDER AN INTERSTATE COMPACT. *Edward J. Cleary*. 1967. 352 pages. $10.00 cloth, $2.95 paper.

THE PESTICIDE PROBLEM: AN ECONOMIC APPROACH TO PUBLIC POLICY. *J. C. Headley and J. N. Lewis*. 1967. 160 pages. $3.50 paper.

WATER RESEARCH. *Allen V. Kneese and Stephen C. Smith, eds.* 1967. 534 pages. $15.00.

ENVIRONMENTAL QUALITY IN A GROWING ECONOMY: ESSAYS FROM THE 1966 RFF FORUM. *Henry Jarrett, ed.* 1966. 188 pages. $5.00.

WATER DEMAND FOR STEAM ELECTRIC GENERATION: AN ECONOMIC PROJECTION MODEL. *Paul H. Cootner and George O. G. Löf*. 1966. 156 pages. $4.00 paper.

QUALITY OF THE ENVIRONMENT: AN ECONOMIC APPROACH TO SOME PROBLEMS IN USING LAND, WATER, AND AIR. *Orris C. Herfindahl and Allen V. Kneese*. 1965. 104 pages. $2.00 paper.

THE ECONOMIC DEMAND FOR IRRIGATED ACREAGE: NEW METHODOLOGY AND SOME PRELIMINARY PROJECTIONS, 1954–1980. *Vernon W. Ruttan*. 1965. 154 pages. $5.00.

WATER POLLUTION: ECONOMIC ASPECTS AND RESEARCH NEEDS. *Allen V. Kneese*. 1962. 116 pages. $1.75 paper.

Books from Resources for the Future, Inc. may be ordered from bookstores or from The Johns Hopkins Press, Baltimore, Maryland 21218.

ISBN 0-8018-1215-1